Eigenvalues of the Laplacian for Hecke Triangle Groups

Recent Titles in This Series

(*Continued in the back of this publication*)

MEMOIRS
of the
American Mathematical Society

Number 469

Eigenvalues of the Laplacian
for Hecke Triangle Groups

Dennis A. Hejhal

May 1992 • Volume 97 • Number 469 (end of volume) • ISSN 0065-9266

American Mathematical Society
Providence, Rhode Island

1991 *Mathematics Subject Classification.*
Primary 11F30, 11F72, 30F35, 35P05, 65N35.

Library of Congress Cataloging-in-Publication Data

Hejhal, Dennis A.
  Eigenvalues of the Laplacian for Hecke triangle groups/Dennis A. Hejhal.
  p. cm. – (Memoirs of the American Mathematical Society, ISSN 0065-9266; no. 469)
  Includes bibliographical references.
  ISBN 0-8218-2529-1
  1. Selberg trace formula. 2. Automorphic functions. 3. Eigenvalues. 4. Laplacian operator.
I. Title: Hecke triangle groups. II. Series.
QA241.H38 1992   92-6943
[512′.7]–dc20   CIP

Memoirs of the American Mathematical Society

This journal is devoted entirely to research in pure and applied mathematics.

Subscription information. The 1992 subscription begins with Number 459 and consists of six mailings, each containing one or more numbers. Subscription prices for 1992 are $292 list, $234 institutional member. A late charge of 10% of the subscription price will be imposed on orders received from nonmembers after January 1 of the subscription year. Subscribers outside the United States and India must pay a postage surcharge of $25; subscribers in India must pay a postage surcharge of $43. Expedited delivery to destinations in North America $30; elsewhere $82. Each number may be ordered separately; *please specify number* when ordering an individual number. For prices and titles of recently released numbers, see the New Publications sections of the *Notices of the American Mathematical Society.*
  Back number information. For back issues see the *AMS Catalogue of Publications.*
  Subscriptions and orders should be addressed to the American Mathematical Society, P. O. Box 1571, Annex Station, Providence, RI 02901-1571. *All orders must be accompanied by payment.* Other correspondence should be addressed to Box 6248, Providence, RI 02940-6248.
  Copying and reprinting. Individual readers of this publication, and nonprofit libraries acting for them, are permitted to make fair use of the material, such as to copy a chapter for use in teaching or research. Permission is granted to quote brief passages from this publication in reviews, provided the customary acknowledgement of the source is given.
  Republication, systematic copying, or multiple reproduction of any material in this publication (including abstracts) is permitted only under license from the American Mathematical Society. Requests for such permission should be addressed to the Manager of Editorial Services, American Mathematical Society, P. O. Box 6248, Providence, RI 02940-6248.
  The owner consents to copying beyond that permitted by Sections 107 or 108 of the U.S. Copyright Law, provided that a fee of $1.00 plus $.25 per page for each copy be paid directly to the Copyright Clearance Center, Inc., 27 Congress Street, Salem, MA 01970. When paying this fee please use the code 0065-9266/92 to refer to this publication. This consent does not extend to other kinds of copying, such as copying for general distribution, for advertising or promotion purposes, for creating new collective works, or for resale.

Memoirs of the American Mathematical Society is published bimonthly (each volume consisting usually of more than one number) by the American Mathematical Society at 201 Charles Street, Providence, RI 02904-2213. Second-class postage paid at Providence, Rhode Island. Postmaster: Send address changes to Memoirs, American Mathematical Society, P. O. Box 6248, Providence, RI 02940-6248.

10 9 8 7 6 5 4 3 2 1   97 96 95 94 93 92

TABLE OF CONTENTS

ABSTRACT

This Memoir is concerned with computational aspects of the Selberg trace formalism. The problem is studied in one of its most basic settings; viz. Schwarz triangle groups with angles $(\frac{\pi}{2}, \frac{\pi}{N}, \frac{\pi}{\infty})$.

In addition to considering the usual type of eigenfunction, our discussion also includes an analysis of *pseudo* cusp forms and their residual effects.

Key words and phrases.

Fuchsian group, Selberg trace formula, eigenvalue of the Laplacian, Eisenstein series, cusp form, pseudo cusp form, oldform, newform, Fourier coefficient, Hecke operator, multiplicative relation, conditioning-level.

O.   FOREWORD

The theme of this Memoir can be loosely described as *computational aspects* of the Selberg trace formalism.

Over the past 10 years or so, this area has begun to blossom into an increasingly important subdiscipline (of the greater TF).

There are several reasons for this.

One reason is applications: numerical results involving the TF can often be applied to *other* areas -- areas ranging anywhere from analytic number theory and automorphic forms to string theory and quantum chaos. Applications of this kind provide one with an important incentive.

The other reasons are more intrinsic.

To appreciate the overall situation, bear in mind what every student of the TF knows; namely that several of the TF's most basic (and sought-after) objects tend to be *highly* transcendental -- and not really "accessible" by way of any type of explicit formula.

This causes serious problems.

In situations like this, passing to an *alternate* perspective will occasionally prove useful.

The purpose of a computational approach to the STF is to try to make the "inaccessible" objects more "tangible" -- by computing them to, say, *10* decimal places. One can then seek to explore these objects further by employing the "time-honored technique" of "simply looking at them."

The ultimate hope [in all this] would be to achieve a rich interplay between theory and experiment.

To some extent at least, this interplay *does* occur.

Significant progress on the various experimental fronts has been made possible, in large part, due to the availability of supercomputers like the CRAY-2 and CRAY-YMP.

One of our primary aims in this Memoir will be to provide readers with some "feel" for the *type* [and extent] of progress which can be expected from such machines using relatively straightforward Fortran codes.

The Memoir will consist of 2 chapters.

The first, and principal, one deals with the spectral theory of the (non-Euclidean) Laplacian Δ over Hecke triangle groups $\mathfrak{G}(2\cos\frac{\pi}{N})$. In addition to considering the *usual* type of eigenvalues and eigenfunctions, it will be seen how the related concepts of pseudo cusp forms, almost cusp forms, and residual effects all enter the picture quite naturally.

Received by the editors September 17, 1990. Revised June 1, 1991.

Besides providing additional support for the (still conjectural) Sarnak-Phillips philosophy, our results *suggest* that pseudo cusp forms $\overline{\Phi}$ having a single logarithmic singularity at $\rho \equiv e^{\pi i/N}$ are distinguished, at least locally, by a certain *extremal* property which is decisive in allowing them to be detected so sharply.

Further investigation along these lines [both theoretically and experimentally] may well show that the space of pseudo cusp forms has a *richer* structure than previously thought.

The second chapter is a slightly revised version of the author's 1988 [Hua Memorial] paper dealing with $PSL(2,\mathbf{Z}) \setminus H$. By including this paper here ([1]), we hope to make the present Memoir more self-contained (and generally readable).

Though experimental, our CRAY code for $PSL(2,\mathbf{Z}) \setminus H$ was successful in allowing one to explore the spectrum of Δ [on $PSL(2,\mathbf{Z})$] out to about $R = 500$ or so . Chapter 2 provides a detailed account of the actual runs -- and associated numerical subtleties.

[In any work based either in whole or in part on the computer, the question of *rigor* inevitably comes up. On this matter, our working philosophy has been strongly influenced by 3 things:
(A) the use to which we intend to put any of our output;
(B) the fact that this whole area is still (mainly) in an exploratory phase;
(C) the limit on available resources.

Under the circumstances, *particularly when the ultimate goal is insight,* it would seem far preferable to get a good grip on *100N* things with probability $1 - \varepsilon$ than on *N* things with probability $1 - \varepsilon^{100}$.

It is this path that we have sought to follow.

Measures are taken along the way to keep ε to a *practical,* if not theoretical, minimum.]

Before turning to the actual mathematics, three final remarks are in order.

First: concerning the scope of chapters 1 and 2, it will (soon) become clear that we have only scratched the tip of an iceberg. Many avenues remain wide open for further study. It is hoped that this Memoir may stimulate additional activity in these areas.

Secondly: I would be completely remiss if I didn't stop to acknowledge the important impetus that Hua Loo-Keng provided to these eigenvalue investigations during the years 1979-1985. A significant portion of my outlook (and style) in these matters was shaped by conversations the two of us had.

Finally: it is a pleasant duty to thank Scott Wolpert for his insightful comments on chapter 1. His watchful eye prevented (what would have been) an embarrassing slip in Table 4.

([1]) as a *supplement* to chapter 1

I. EIGENVALUES OF THE LAPLACIAN FOR HECKE TRIANGLE GROUPS

ABSTRACT. The object of this paper is to report on a series of experiments aimed at computing eigenvalues of the Laplacian for a variety of Hecke triangle groups $G(2\cos\frac{\pi}{N})$. [The results we obtain will also shed new light on *pseudo* cusp forms and their residual effects.] (*)

§1. <u>Introduction</u> <u>and</u> <u>Preliminary</u> <u>Remarks</u>. In recent years, much effort has been devoted to studying the Selberg trace formula and associated spectral decomposition for $L_2(\Gamma \backslash H)$ in the case where Γ is a cofinite Fuchsian group. Many very substantial advances have been made on the theoretical front. The *numerical* side has generally tended, however, to lag considerably behind.

To this day: the *explicit* computation of eigenfunctions of the Laplacian for $\Gamma \backslash H$ remains a largely *un*explored aspect of the STF. This deficiency seems to be one well worth rectifying. In fact: the development of efficient algorithms in this area would not only be of interest intrinsically, but could (also) be expected to shed new light on any number of sticky/theoretical issues surrounding the TF itself.

At present: matters can safely be said to be in an exploratory stage at best.

Under the circumstances, it seems reasonable to (temporarily) dispense with the pursuit of totally rigorous/error-controlled algorithms in favor of techniques which are a bit looser -- or even experimental.

Cf. [2],[13,Appendix C],[19],[41] for a quick look at some of the existing approaches.

In a recent paper [14], we discussed a new algorithm ([1]) which allowed one to "explore" the discrete spectrum of $PSL(2,Z) \backslash H$ relatively painlessly all the way out to $\lambda = 250000$ or so. The method required access to a machine like an XMP or CRAY2 and a willingness to (ultimately) spend something like 30 cpu hours testing a λ - interval of length 500. The corresponding time at $\lambda = 62500$ is about 7 hrs. [Both times can probably be reduced by a factor of $3 \sim 5$ simply by restructuring the code; bear in mind *too* that the number of useful Fourier coefficients grows like $\sqrt{\lambda}$.]

(*) With support from NSF Grant DMS 89-10744 and computer time from the Minnesota Supercomputer Institute (CRAY2 , XMP416), Pittsburgh Supercomputing Center (YMP832), and San Diego Supercomputer Center (XMP48).

([1]) still based on collocation 〈 Note that paper [14] ≡ chapter 2 . 〉

The algorithm in [14] proceeds by having the machine search for those "candidates" (in the basic Ansatz) which satisfy a variety of numerical criteria *consistent* with being a true eigenfunction. As such: the (over-all) strategy is basically one of achieving high probability -- as opposed to complete certainty.

⟦This "outlook" is nothing new. Hands-on numerical work often involves extrapolations of this kind. In some sense: the *main* difficulty is "arranging things" so that the resulting probability is *sufficiently* close to 1 .⟧

It is natural to ask whether ideas similar to [14] can be profitably applied to groups *other* than PSL(2,**Z**) .

One *conjectures* that the answer is strongly affirmative. (²)

This Note is the first installment in (what we hope will be) a series of several papers dealing with this matter. (³)

It is reasonable to restrict oneself at the outset to the case of weight 0 and trivial multiplier system. We shall also require that Γ have at least 1 cusp. (Set $\lambda \equiv \frac{1}{4} + R^2$, as usual.)

One of the simplest generalizations of PSL(2,**Z**) is the Hecke group $\mathfrak{G}(2\cos\frac{\pi}{N})$. Cf. [12,pp.592,629] . The bulk of *this* paper will focus on $\mathfrak{G}(2\cos\frac{\pi}{N})$ for N = 4,5, 6,7 .

For the sake of clarity, we stop to recall a few basic properties of $\mathfrak{G}(2\cos\frac{\pi}{N})$. First and foremost: $\mathfrak{G}_N \equiv G(2\cos\frac{\pi}{N})$ is generated by

$$E(z) = -\frac{1}{z} \quad \text{and} \quad T(z) = z + \mathcal{L}$$

where $\mathcal{L} = 2\cos(\frac{\pi}{N})$. The number N is a positive integer $\geqq 3$. It is easily seen that

$$\mathcal{F}_N = H \cap \{|z| > 1\} \cap \{|Re(z)| < \frac{\mathcal{L}}{2}\}$$

is a fundamental region for $\mathfrak{G}_N \setminus H$ and that $\mathfrak{G}_N$ has signature

---

(²) In this regard: it is interesting to observe that the main ideas in [14] are vaguely reminiscent of A.Selberg's 1966 proof for the analytic continuation of the Eisenstein series. Cf. [13,pp.711,719(9),720(top)].

(³) Though, in this installment, we focus mainly on some special groups, the numerical groundwork we lay can clearly be extended to more general groups Γ . The subtleties that appear in the *special* cases can be expected to *recur* ⟦possibly with a vengeance⟧ when more general Γ are used. For this reason, the simpler groups actually serve as as an important testing-ground.

$$(g, n ; \tau_1, \ldots, \tau_n) = (0, 3 ; 2, N, \infty) \quad ,$$

Cf. [13,pp.5,569],[24,pp.227,235],[12,pp.609-616] . The group $\mathfrak{C}_N$ is (thus) a particular realization of the Schwarz triangle group $T(\frac{\pi}{2}, \frac{\pi}{N}, \frac{\pi}{\infty})$. It virtually goes without saying here that $\mathfrak{C}_3 = PSL(2,\mathbf{Z})$ and that $\mathfrak{C}_N$ admits an obvious symmetry with respect to the imaginary axis.

The group $\mathfrak{C}_N$ is known to be *commensurable* with $PSL(2,\mathbf{Z})$ iff $N = 3,4,6$. Cf. [21] and [25,42] for the "if" and "only if", respectively.

Our *goal* is to study the discrete spectrum of $\mathfrak{C}_N \setminus H$ for a variety of N .

The results we obtain (in §§5-6) will serve to both amplify and extend some earlier work by A.Winkler [45] .

Winkler's approach is substantially different than ours.

Prior to discussing the "mechanics" of *our* algorithm, it is useful to step back just a bit -- and make several remarks pertaining to a very important conjecture (1985) of P.Sarnak and R.S.Phillips.

To this end, let Γ be any cofinite Fuchsian group with $h > 0$ inequivalent cusps. Let $E_j(z;s)$ be the usual Eisenstein series associated with the jth cusp. The spectral decomposition of a generic $f \in L_2(\Gamma \setminus H)$ will then look like

$$(1.1) \qquad f \sim \sum_n c_n \varphi_n + \sum_{j=1}^{h} \int_0^{\infty} g_j(t) E_j(z; \tfrac{1}{2} + it) \, dt \quad .$$

Cf. [13,pp.91,291] .

In a computational approach to the STF, one would want to be able to calculate the φ_n & E_j quite explicitly. With present technology, this desire remains largely a distant (but very tantalizing!) dream.

For *arithmetic* groups Γ , it is known that there exist infinitely many φ_n [with $0 = \lambda_0 < \lambda_1 \leq \lambda_2 \leq \ldots \rightarrow \infty$] and that the E_j can effectively be regarded as Epstein zeta functions (a type of zeta function long familiar in number theory). On the latter score: cf., eg , [39,pp.4,23,54,70], [27,pp.293,294,286] , and [37,§9] .

For *nonarithmetic* Γ , the situation changes drastically. There the E_j are rather inexplicit -- being defined along $\{ Re(s) = \frac{1}{2} \}$ only by way of analytic continuation (wrt s). This "intangibility" has a serious impact on Weyl's law ([13,p.210]) .

In fact: according to the conjecture of Sarnak and Phillips [29,30] , one should *expect* -- in the absence of any kind of algebraic symmetry (or grouptheoretic inclusion) on Γ -- that the *total* number of linearly independent φ_n is *finite* (not infinite!!).

The original conjecture of Sarnak and Phillips is now viewed as part of a much larger philosophy [34]. Within the latter: there is *no* difficulty accomodating (nonarithmetic) groups with symmetries.

In the case of $\mathfrak{C}_N$, recall that there was a symmetry with respect to the imaginary axis. Exactly as in [13,p.590(13)] , one finds that the spectral decomposition of $L_2(\mathfrak{C}_N \setminus H)$ splits into two "halves", one "even" and one "odd." The Eisenstein series appears *only* in the "even" half.

The *odd* portion of $L_2(\mathfrak{C}_N \setminus H)$ will therefore be purely discrete. There is no difficulty obtaining Weyl's law for this half. Cf. [43,§§ 6.5, 6.7] and [35,pp. 69,72(†)] .

The Sarnak-Phillips philosophy refers mainly to the *other* (i.e. "even") half of L_2 .

Since the first eigenvalue of any triangle group is automatically bigger than $\frac{1}{4}$ (cf. [13,p.583(8)]), the relevant conjecture can be stated as follows:

$$(\bigstar) \qquad \left\{ \begin{array}{l} \text{for } N \neq 3,4,6 \text{ , the Hecke group } \mathfrak{C}(2\cos\frac{\pi}{N}) \\[2ex] \text{should admit } no \text{ even cusp forms} \end{array} \right\} \quad .$$

The contrast between odd/even and arithmetic/nonarithmetic is indeed striking. It is now apparent why $\mathfrak{C}_N$ is such a natural candidate for the computer.

Trying to test the Sarnak-Phillips conjecture for a "random" Γ is never easy. There are many difficulties. The most serious one (perhaps) is recognizing a "true" φ_n when one "sees" one. To explain this point, assume for a moment that Φ is a cusp form on $\Gamma_o \setminus H$ (with eigenvalue $\frac{1}{4} + R^2$ and multiplicity 1) which is *known* to be "destroyed" under some real analytic Teichmüller-style deformation Γ_ε of Γ_o . Cf. [29,30,4]. The groups Γ_ε are all Fuchsian with h inequivalent cusps. To keep things simple, take h = 1 . By [13,pp.231-232,143-148] , it follows that $E_\varepsilon(z;s)$ has a simple pole $\frac{1}{2} - \eta_\varepsilon + iR_\varepsilon$ very close to $\frac{1}{2} + iR$ (as $\varepsilon \to 0$). [4] Since $E_o(z;s)$ can't have a pole at $\frac{1}{2} + iR$, something fishy must happen in the limit. In §9(G), we'll examine the underlying formalism. The upshot of this paradigm is that

$$(1.2) \qquad \sqrt{\eta_\varepsilon}\; E_\varepsilon(z;\tfrac{1}{2}+iR_\varepsilon)$$

will tend to look more & more like (constant)Φ as $\varepsilon \to 0$. From a *numerical* standpoint, then, once ε drops below a certain threshold, it will be *impossible* to

---

[4] Of course: $\eta_\varepsilon > 0$.

tell the difference between (1.2) and a true cusp form on $\Gamma_\varepsilon \setminus H$. [5]

It's now the old problem... How does one tell (on a computer) when a *real* number η is exactly zero??

For obvious reasons, matters shift to ε and it becomes clear that an efficient "tracking mechanism" for $\frac{1}{2} - \eta_\varepsilon + iR_\varepsilon$ would be a real asset. (One presumes here that everything in sight depends real analytically on ε.)

We hope to return to this "tracking" idea in a subsequent paper. For now, however, we prefer to return to $\mathfrak{C}_N$ -- and exploit one of its intrinsic advantages. Namely: since $\mathfrak{C}_N$ is a triangle group, the Teichmüller space of $\mathfrak{C}_N$ is zero-dimensional [3]. Hence: *no* nontrivial deformations of $\mathfrak{C}_N$ exist.

In other words: things are *rigid*.

This makes it tempting to believe that, at least for small N (and *bounded* R), we are *not* going to be faced with the occurrence of any "phonies" similar to (1.2) in our actual computer runs. ⟦One can only hope for the best. If such objects do occur, conjecture (✿) could be in big trouble -- epistemologically.⟧

The reason for taking N small is very simple. It is known that the poles of $E(z;s;\mathfrak{C}_N)$ become everywhere dense along $\{Re(s) = \frac{1}{2}\}$ whenever $N \longrightarrow \infty$. Cf. [13, p.579].

Similar concerns enter the "picture" with regard to R. In fact: by applying [13,pp.210(i), 456(2.22), 476(3.7)], it is easily seen that η will typically look something like $c\frac{\log R}{R}$ (on average) for each $N \neq 3,4,6$ ⟦*at least* if (✿) holds⟧.

These remarks show that, even though $\mathfrak{C}_N$ is "virtually trivial" group theoretically, a certain amount of caution must still be exercised.

In short: theory and experiment need to go hand-in-hand.

§2. The Procedure in a Nutshell. Our aim is to find cusp forms $\varphi(z)$ for $\mathfrak{C}(2\cos\frac{\pi}{N})$. By virtue of an earlier remark, we already know λ must be strictly greater than $\frac{1}{4}$. This leads to $R > 0$ and a Fourier expansion

(2.1)
$$\varphi(x+iy) = \sum_{n=1}^{\infty} c_n y^{\frac{1}{2}} K_{iR}\left(\frac{2\pi n y}{\ell}\right) \left\{ \begin{array}{c} \cos\left(\frac{2\pi n x}{\ell}\right) \\ \text{------------} \\ \sin\left(\frac{2\pi n x}{\ell}\right) \end{array} \right\}$$

depending on whether φ is even or odd. As usual: $\lambda = \frac{1}{4} + R^2$. There is no loss of generality in assuming that $c_n \in \mathbb{R}$.

The RHS of (2.1) is automatically invariant under $z \longmapsto z + \ell$. To achieve

---

[5] Bear in mind here that the Fourier coefficients in the 0-order term for (1.2) tend to 0 as $\varepsilon \longrightarrow 0$.

full automorphy, we need to ensure that

(2.2) $\varphi\left(-\frac{1}{z}\right) \equiv \varphi(z)$.

For a "true" cusp form, we stress that (2.1) must be absolutely convergent on *all* of H . Cf. [13,pp.23-25] .

The K-Bessel function is defined via

(2.3) $K_{iR}(X) = \frac{1}{2} \int_{-\infty}^{\infty} e^{-X\cosh t} e^{iRt} \, dt$.

Due to the extremely small size of $K_{iR}(X)$ for $R > 20$, it is best ([14,17]) to compute (2.3) by *bending* the contour in a manner similar to stationary phase. In this way: there is no difficulty calculating $\exp(\frac{\pi}{2}R)K_{iR}(X)$ to 10 or 11 places for R-values all the way out to 75000.

The algorithm we use in connection with (2.1)+(2.2) is very similar to the one found in [14] . There is one major difference, however. Namely: the group $\mathfrak{G}_N$ does not generally admit any Hecke operators. Cf. [25] , [36,§4] , [38]. This effectively eliminates any hope of determining R by use of some sort of multiplicative relations among the c_n .

To circumvent this difficulty, we proceed as follows. First of all: *recall that* (2.2) is equivalent to

(2.4) $\sum_{n=1}^{\infty} c_n I_n(z, R) = 0$ $(^1)$.

We now select two batches of points $\{z_1, \ldots, z_{M-1}\}$ & $\{w_1, \ldots, w_{M-1}\}$ in $\mathfrak{F}_N$ (with suitable M) and repeatedly solve

(2.5') $\sum_{n=2}^{M} c_n I_n(z_j, R) = -I_1(z_j, R)$, $1 \leq j \leq M-1$;

(2.5'') $\sum_{n=2}^{M} c_n I_n(w_j, R) = -I_1(w_j, R)$, $1 \leq j \leq M-1$.

The *goal* is to determine those R-values for which the solution sets $(c_2', c_3', \ldots, c_M')$ and $(c_2'', c_3'', \ldots, c_M'')$ match [as far as possible] .

This is the *new strategy* in a nutshell.

$(^1)$ And that the terms I_n *ultimately* decay exponentially fast wrt n . [To bring the first few I_n "up" closer to 1 , it is customary to premultiply $K_{iR}(X)$ by $\exp(\frac{\pi}{2}R)$. We *tacitly* assume that this has been done. See [7] for the relevant asymptotics of $K_{iR}(X)$.]

There is no difficulty incorporating this change in our earlier CRAY code. (See appendix A for a sample program.)

For the sake of completeness, we remind the reader that there are 3 length-scales to be concerned about. In short:

H1 = the "coarse" grid on which the initial computation of all the K-Bessel functions takes place ;

H2 = the "finer" scale on which [by Lagrange interpolation] the matrices $(2.5')(2.5'')$ are all assembled and then manipulated to produce *tentative* R-values ;

H3 = the "final" scale on which "blow-ups" are made around each of the tentative R-values .

See [14] for an analysis of (some of) the numerical subtleties.

In general: we take H3 to be 10^{-6} and H2 to be no more than 1% of $\frac{4\pi}{AR}$, where

$$A = \mu(\mathcal{F}_N) = \pi\left(1 - \frac{2}{N}\right) = \text{the hyperbolic area of } \mathcal{F}_N .$$

The number $\frac{4\pi}{AR}$ corresponds to the *average distance* between successive R-values assuming the validity of Weyl's law. Cf. [43, thms 6.5.5 & 6.7.1].

— — —

Two additional remarks (pertaining to program structure) should also be mentioned here.

In our earlier program: recall that *changes-of-sign* in

$$c_4 + 1 - c_2^2$$

were used to help detect "promising" H2 intervals. This aspect of the code is now replaced by a simple IF- statement involving

$$sgn\left(c_2' - c_2'' , c_3' - c_3'' , c_4' - c_4''\right) .$$

With regard to H3 , we remark that the final R-values are obtained by *minimizing* a functional rather like:

$$|c_2' - c_2''| + |c_3' - c_3''| + |c_4' - c_4''| .$$

It is *not* feasible to actually solve $\{c_2' = c_2'' , c_3' = c_3'' , c_4' = c_4''\}$ directly. Compare [14,§4] .

[Note that this technique does not automatically force the *higher* differences $|c_k' - c_k''|$ to be small. Their smallness is a useful form of "insurance" (and an *excellent* measure of overall accuracy)...]

The foregoing ideas are all implemented in a standard Cray-Fortran code (having length $\approx$ 1200 lines). Only single-precision variables are used...

To solve (2.5), we use standard Gauss-elimination. Cf. [9,pp.65-72(II)].

§3. Some Theoretical Difficulties.   The remarks in §2 were only a brief outline; any number of difficulties were simply glossed over. It is *essential* to clarify several of the more important ones before going any further.

In the discussion that follows, we'll concentrate mainly on theoretical matters. The reader is referred to [14] for a fuller discussion of the computational side.

Our remarks will be largely heuristic.

Concern #1 is the matter of an *implied* normalization. We have obviously assumed in (2.5) that

$$(3.1) \qquad\qquad c_1 = 1 \qquad .$$

Cusp forms satisfying (3.1) will be referred to as "unit normalized." Insofar as we are not dealing with any kind of multiple eigenvalue, this normalization seems perfectly legitimate. Things happen for *reasons* ; it is difficult to imagine what $c_1 = 0$ could possibly *mean* (especially in a *nonarithmetic* setting).

Bear in mind here that $\mathfrak{C}_N$ is a *maximal* Fuchsian group. Cf. [10,28,40]. As such: its normalizer can't be something strictly bigger. This effectively rules out any kind of *intrinsic* (or representation-theoretic) reason for multiple eigenvalues. Compare [31]. ([1])

Concern #2 is the possibility of finding two eigenvalues unusually close together. Near such R-values, it seems virtually certain that (2.5) must encounter a numerical "singularity" (*or* conditioning disaster). To penetrate the "static", some sort of refined algorithm will obviously be necessary.

In our current approach, we decided to simply ignore this possibility and hope for the best.

([1]) There is, of course, *another* possibility to contend with here (particularly from the standpoint of computation). Namely: c_1 might simply be *exceptionally* small compared to the prevailing size of its immediate c_k-neighbors.

An imbalance of this type will generally tend to create serious difficulties in (2.5). The problem stems from the fact that a unit normalization has been imposed in (2.5). Taking $c_1 \equiv 1$ *effectively* makes the tail $\sum_{M+1}^{\infty} c_k I_k(z,R)$ that much larger, possibly to the point of "overpowering" $I_1(z_j,R)$ & $I_1(w_j,R)$ in (2.5'),(2.5"). Compare (5.1). This (in turn) can easily lead to an overall degradation in $|c_k' - c_k''|$ quality. Depending on the extent of this, the number R_n can wind up getting "completely lost."

This danger can clearly be mitigated by using alternate normalizations (such as $c_2 \equiv 1$) -- and by testing a variety of M-values.

For *small* values of R and N, it is reasonable to expect that (2.5) will encounter only minimal difficulties of the preceding kind.

At the same time: *note* that the existence of Hecke operators typically carries with it some type of apriori bound for c_k/c_1 . Cf. [38,pp.80,90(top)] and §4 . The aforementioned difficulties can therefore occur *only if* $\mathfrak{C}_N$ is nonarithmetic.

Given the structure of our code, there can be little doubt but that pairs of "nearly multiple" eigenvalues become "invisible" once their gap falls below a certain δ . With H2 at 1% , a threshold in the neighborhood of 5% [of $4\pi/(AR)$] seems reasonable. To achieve a better "resolution", H2 can (of course) be provisionally reduced. [2]

Since the largest R-value in this paper is only about 80, this matter does not seem overly pressing (at least for now).

Concern #3 is the possibility of encountering something like (1.2). By their very nature: such objects (can) only occur in connection with the *even* "half" of L_2 . For arithmetic $\mathfrak{C}_N$, we certainly don't expect any problems [3]. Cf. [39,pp. 23,54] , [13,pp.76,549] , [27,p.294(32)] .

In *non*arithmetic cases, we can only hope for the best. Cf. (✗) and the last few paragraphs of §1.

There is one *final* concern that's a mixture of theory and numerics. Namely: the possibility of encountering objects similar to the <u>pseudo</u> cusp forms defined in [16].

This possibility is a very serious matter affecting (as we shall soon see) *both* the even and odd "halves" of L_2 .

To put things in the proper perspective, we first recall that any sensible algorithm must observe 3 main precautions:
(i) it is essential that the *same* R-values be obtained even when the points z_j & w_j are varied ;
(ii) likewise for the (associated) c_n ;
(iii) the (purported) c_n should not grow "too rapidly" as n increases.

These precautions can [and should] be implemented *only* so far as the basic numerical configuration allows. Cf. [14] near (2.7). Precautions (i) and (ii) are self-evident; the problems begin with (iii).

Item (iii) is included to "make sure" that (2.1) converges on *all* of H . The difficulty occurs in trying to make (iii) more explicit. For cuspidal φ , the procedure in [13,pp.585,587] yields $c_n = O(n^{1/2})$ easily enough, but the size of the implied constant is *not* very precise. [Remember that we want φ to be *unit-normalized*!]

When $\mathfrak{C}_N$ lacks Hecke operators, there is *no* analog of the Ramanujan-Petersson conjecture to serve as a guide -- and it is not at all clear what to do. Looking at a *finite* number of c_n is not sufficient to determine the true region of convergence of (2.1) ...

This leads to difficulties which are potentially disastrous.

To see this: suppose that the test points $\{z_j\}$ & $\{w_j\}$ were not chosen very wisely. E.G. suppose that the sets $\{E(z_j)\}$ & $\{E(w_j)\}$ never drop below y =

---

[2] Note that *similar* difficulties occur even in studying the zeros of $\mathfrak{Z}(s)$. Cf. [6,23, 51]. (The question of extending [51] to the case of the *Selberg* zeta function of $\mathfrak{C}_N \backslash H$ is largely moot -- since no good means of computing the latter is known.)

[3] Assuming, say, GRH.

$\sin(\frac{\pi}{N})$. We can then define various types of automorphic eigenfunctions Φ having *singularities* in a tiny neighborhood of $\rho \equiv \exp(\pi i/N)$ [and polynomial growth at $i\infty$]. The presence of (nonremovable) singularities implies that the standard Fourier expansion for Φ [with K-Bessel functions] *cannot* converge below $y = \sin(\frac{\pi}{N})$. However, it *will* converge at z_j , w_j , $E(z_j)$, $E(w_j)$. Bear in mind here that the $\mathfrak{C}_N$ - orbit of ρ never rises above $y = \sin(\frac{\pi}{N})$. Cf. [25,p.201(middle)] .

The essential point is now this. If the *zero-order* term in Φ's Fourier expansion is 0 (or even just sufficiently small), there is a *chance* that Φ will be detected by (2.5). The *deciding factor* is whether the truncation we made (after M terms) is still reasonably accurate at z_j , w_j , $E(z_j)$, $E(w_j)$.

Note that it is *not* necessarily true here that precaution (iii) must fail. The first few c_n could well be abnormally small [for some reason *consistent* with having having a small tail at z_j , w_j , $E(z_j)$, $E(w_j)$].

In general, of course, one expects $|c_n|$ to look something like

(3.2) $$\exp\left[\frac{2\pi n}{\ell} \operatorname{Im}(\rho)\right] \qquad .$$

By adjusting the singularity, we can make our original Φ be either even or odd. The standard example (here) is:

(3.3) $$\sum_{i=1}^{Q} a_i\, G_s(z; \tau_i) \qquad , \qquad a_i \in \mathbb{R} \qquad .$$

One can also form various τ- derivatives as in [8] . The function G_s is the *usual* Green's function for $\mathfrak{C}_N \backslash H$. Cf. [13,pp.33,250(bot)] .

By virtue of [13,p.42(6.8)] , the zero-order term for (3.3) is

(3.4) $$\left[\sum_{i=1}^{Q} a_i\, E(\tau_i ; s)\right] \frac{y^{1-s}}{1-2s} \qquad .$$

A little experimentation shows that the bracket can be made to vanish at <u>any</u> pre-assigned point of $\{\operatorname{Re}(s) = \frac{1}{2}\}$ by proper choice of a_i & τ_i . In fact: $Q = 2$ will suffice. (Reality concerns are most easily addressed by means of the functional equation. An *alternate approach* is to premultiply $E(w;s)$ by $A(s)$ as in [13, p.131].)

In view of examples of this kind: it becomes rather clear that one should use batches $\{z_j\}$ & $\{w_j\}$ whose E-images lie *strictly below* $y = \sin(\frac{\pi}{N})$.

Various hybrid batches can also be considered, but the essential feature remains the same : one must try to use the *geometry* to discourage any putative φ from having singularities in $\mathcal{F}_N \cup \partial \mathcal{F}_N$.

We are tempted to summarize these remarks with the phrase

"forewarned is forearmed."

§4. <u>Coefficient Relations for $N = 4$ and 6</u>.   Since $\mathfrak{C}_4$ and $\mathfrak{C}_6$ are commensurable with $PSL(2,\mathbf{Z})$, it is reasonable to *conjecture* that some type of Hecke operators will (now) exist -- and that the Fourier coefficients of any unit-normalized φ will automatically satisfy certain multiplicative relations (at least if λ has multiplicity 1).

To treat $N = 4$ and 6 simultaneously, we set

$$q = \frac{N}{2}$$

and (then) observe that $\mathcal{L} = 2\cos(\frac{\pi}{N}) = \sqrt{q}$.

Let $\mathcal{Y}$ be the subgroup of $PSL(2,\mathbb{R})$ which is generated by

$$w \longmapsto w+1 \qquad and \qquad w \longmapsto -\frac{1}{qw} \qquad .$$

The group $\mathcal{Y}$ is nothing but $\mathfrak{C}_N$ viewed under the auxiliary mapping

(4.1) $z = \sqrt{q}\, w$.

We also set:

$$\Gamma_0(q) = \left\{ \begin{pmatrix} a & b \\ c & d \end{pmatrix} \in SL(2,\mathbf{Z}) : c \equiv 0 \mod q \right\} \qquad .$$

The discussion in [21] shows that $\Gamma_0(q) \lesseqgtr \mathcal{Y}$ and that the index is 2. We already know that $\mathcal{Y}$ is a *maximal* Fuchsian group. Cf. §3. The analysis in [1, p.139] immediately implies that:

$$\mathcal{Y} = \text{the normalizer of } \Gamma_0(q) \qquad .$$

With these items in place, it is *now* possible to derive an important connection between cusp forms on $PSL(2,\mathbf{Z})$, $\mathfrak{C}_N$, and $\Gamma_0(q)$.

To explain things, we assume that the reader already has some familiarity with the Atkin-Lehner formalism [1] *and* is willing to grant that similar things should hold for nonholomorphic cusp forms. Compare [26,32] .

In the remarks that follow, we restrict ourselves to the case of "even" forms. The "odd" case is similar.

To get started: let $f_0(z)$ be any Hecke-normalized cusp form on $PSL(2,\mathbb{Z})$ with eigenvalue $\lambda \equiv \frac{1}{4} + R^2$. We therefore have

$$(4.2) \qquad f_0(x+iy) = \sum_{n=1}^{\infty} a_n y^{\frac{1}{2}} K_{iR}(2\pi n y) \cos(2\pi n x)$$

and

$$(4.3) \qquad \sum_{n=1}^{\infty} \frac{a_n}{n^s} = \prod_p \frac{1}{1 - a_p p^{-s} + p^{-2s}} \qquad \bullet$$

Cf. [14, equation (2.3)] .

By an old remark of Rausenberger [33] , the function

$$(4.4) \qquad g_0(z) = f_0(z\sqrt{q}) + f_0\left(\frac{z}{\sqrt{q}}\right)$$

is automorphic (hence *cuspidal*) on $\mathbb{C}_N$. A trivial manipulation yields:

$$(4.5) \qquad g_0(z) = \left(\frac{1}{\sqrt{q}}\right)^{1/2} \sum_{n=1}^{\infty} c_n y^{\frac{1}{2}} K_{iR}\left(\frac{2\pi n y}{\sqrt{q}}\right) \cos\left(\frac{2\pi n x}{\sqrt{q}}\right)$$

where

$$(4.6) \qquad c_n = a_n + \sqrt{q}\, a_{\frac{n}{q}}$$

The symbol $a_{n/q}$ is understood to be 0 if $q \nmid n$.

By applying $z = \sqrt{q}\, w$, we see that

$$(4.7) \qquad h_0(w) = f_0(qw) + f_0(w)$$

is cuspidal on $\mathcal{Y}$. Since f_0 "lives" on $PSL(2,\mathbb{Z})$, the function h_0 is an *oldform* on $\Gamma_0(q)$. Cf. [1,pp.145-146] .

The Fourier expansion of $h_0(w)$ is simply

$$(4.8) \qquad h_0(u+iv) = \sum_{n=1}^{\infty} c_n v^{\frac{1}{2}} K_{iR}(2\pi n v) \cos(2\pi n u) \qquad \bullet$$

This expansion is augmented by the relation:

$$(4.9) \qquad h_0\left(-\frac{1}{qw}\right) = h_0(w) \qquad \bullet$$

We now turn matters completely around and *begin* with any *newform* h(w)
on $\Gamma_0(q) \setminus H$. Cf. [1,p.145] and [32,pp.321-323] .

There are two types of newforms depending on whether

$$h\left(-\frac{1}{qw}\right) = \pm h(w) \qquad ,$$

Cf. [1,p.147] . To obtain automorphy on $\mathcal{Y}$, we obviously want the "+" sign to
hold. Such newforms will be called "proper." (The underlying theme here is
essentially one of *invariant* subspaces. Cf. [1, lemma 25] .)

For proper h , we have:

(4.10)
$$h(w) = \sum_{n=1}^{\infty} c_n v^{\frac{1}{2}} K_{iR}(2\pi n v) \cos(2\pi n u)$$

and

(4.11$_A$)
$$\sum_{n=1}^{\infty} \frac{c_n}{n^s} = \frac{1}{1-c_q q^{-s}} \prod_{p \neq q} \frac{1}{1-c_p p^{-s} + p^{-2s}} \qquad ;$$

(4.11$_B$)
$$c_q = -\frac{1}{\sqrt{q}} \qquad .$$

Cf. [1,pp.147,150] .

By applying $z = \sqrt{q} \, w$, we see that

(4.12)
$$g(z) = h\left(\frac{z}{\sqrt{q}}\right)$$

is cuspidal on $\mathbb{C}_N$ and has Fourier expansion:

(4.13)
$$g(z) = \left(\frac{1}{\sqrt{q}}\right)^{1/2} \sum_{n=1}^{\infty} c_n y^{1/2} K_{iR}\left(\frac{2\pi n y}{\sqrt{q}}\right) \cos\left(\frac{2\pi n x}{\sqrt{q}}\right) \qquad ,$$

It makes sense to call g a *newform* on $\mathbb{C}_N$. The earlier function g_0 will
(then) be called an *old*form.

By abuse of language, we can use the same terminology for cg & cg_0 , $c \neq 0$.

By reviewing the definition of newform in [1,32] , it is easily seen that any two
g and g_0 must be orthogonal (independent of their eigenvalue). In fact:

$$\langle g, g_0 \rangle = \langle h, h_0 \rangle_{\mathcal{Y}} = \frac{1}{[\mathcal{Y} : \Gamma_0(q)]} \langle h, h_0 \rangle_{\Gamma_0(q)} = 0 \qquad .$$

This observation leads to the following reformulation. Let $\lambda = \frac{1}{4} + R^2$ be
any eigenvalue for $\mathbb{C}_N \setminus H$ *with multiplicity* 1 . Let the corresponding unit-
normalized eigenfunction be φ .

If λ is an eigenvalue for PSL(2,$\mathbf{Z}$) , then

$$(4.14_A) \qquad \varphi = (\sqrt{q})^{1/2} g_0$$

for a uniquely determined f_0 on PSL(2,Z) $\backslash$ H . Otherwise,

$$(4.14_B) \qquad \varphi = (\sqrt{q})^{1/2} g(z)$$

for a uniquely determined (proper) newform h on $\Gamma_0(q) \backslash$ H .

All that we're *really* doing here is looking at the oldform/newform decomposition of $\varphi(w\sqrt{q})$ on $\Gamma_0(q) \backslash$ H ...

We won't worry about multiplicity > 1 .

Watching *old*forms appear ⟦on the machine⟧ and verifying ([1]) the *relations* implicit in (4.3), (4.6), (4.11) should prove quite interesting. Especially: (4.11_B).

We summarize things with a diagram:

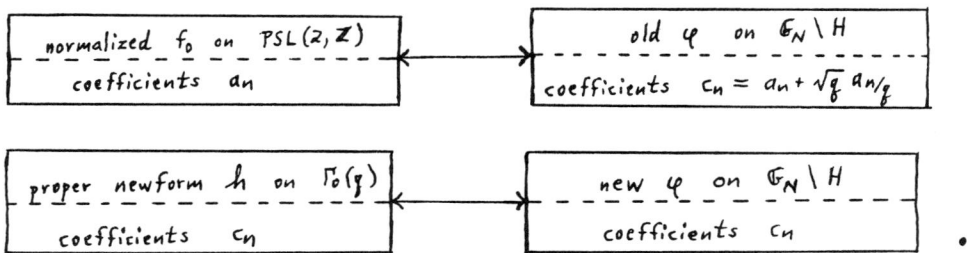

Incidentally: in the case of oldforms, one naturally expects that $\{f_0(w), f_0(qw)\}$ will be a *basis* for the corresponding eigenspace on $\Gamma_0(q)$. The associated multiplicity on $\Gamma_0(q)$ will therefore be 2 . ⟦On $\mathcal{H}$, it'll be *1* .⟧

The corresponding numerology with regard to Weyl's law then goes as follows:

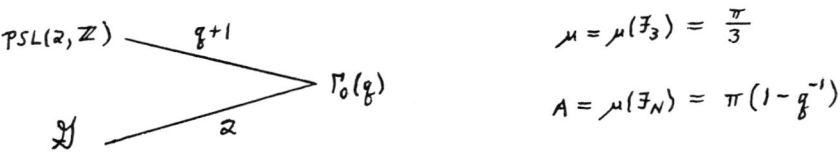

$$\mu = \mu(\mathcal{F}_3) = \frac{\pi}{3}$$

$$A = \mu(\mathcal{F}_N) = \pi\left(1 - q^{-1}\right)$$

(continued)

([1]) for both old and new!

$$p = N[\text{proper newforms with } \lambda_n \leqq X]$$

$$i = N[\text{improper newforms with } \lambda_n \leqq X]$$

$$\frac{(q+1)\mu}{4\pi}X = (\text{multiplicity } 2)\frac{\mu}{4\pi}X + p + i$$

$$\frac{A}{4\pi}X = \frac{\mu}{4\pi}X + p$$

$$\frac{A}{4\pi}X = \frac{\mu}{4\pi}X + i \qquad [\text{by switching } + \text{ to } - \text{ in eq. (4.4)}].$$

Since $(q+1)\mu = 2A$, everything is consistent, and we simply find that:

$$p = \frac{1}{2}(q-1)\frac{\mu}{4\pi}X + [\text{lower order terms}]$$

$$i = \frac{1}{2}(q-1)\frac{\mu}{4\pi}X + [\text{lower order terms}] \qquad .$$

§5. "Odd" Eigenvalues for $N = 4,5,6,7$.   Prior to giving the results, it is useful to say just a few words about the procedure.

Our primary goal [in this set of experiments] was to compute the *odd* eigenvalues of $\mathfrak{C}_4,\ldots,\mathfrak{C}_7$ with $R \leqq 25$ to an R-accuracy of six decimal places.

One wished to do this as efficaciously as possible -- which basically meant that (2.5) had to be "optimally conditioned." This, in turn, meant that some caution had to be exercised in the choice of z_j & w_j .

A batch $\{z_j\} \cup \{w_j\}$ is said to be of *type* $(\alpha_1,\ldots\alpha_r \| \beta_1,\ldots,\beta_s)$ when:

(i) the points $E(z_j)$ are distributed in some regular fashion along the line segments $\{0 \leqq x \leqq \frac{1}{2}\mathcal{L}$, $y = \alpha_i\}$, $1 \leqq i \leqq r$;

(ii) similarly for $E(w_j)$ and $\{0 \leqq x \leqq \frac{1}{2}\mathcal{L}$, $y = \beta_k\}$, $1 \leqq k \leqq s$.

To discourage singularities (as in §3), one requires that:

$$\alpha_i' < \sin\left(\frac{\pi}{N}\right) \quad , \quad \beta_k < \sin\left(\frac{\pi}{N}\right) \quad .$$

Since (2.4) must hold at *any* $z \in H$, it is *not* necessary that the (original) points z_j & w_j lie in $\mathcal{F}_N$. Indeed: for purposes of achieving better conditioning, it would seem *wise* to let $E(z_j)$ & $E(w_j)$ range *all* the way out to $x = \frac{1}{2}\mathcal{X}$. (Intuitively: one wants to spread things out a bit. Cf. figure 1. Several test runs with $N = 6$ convinced us early on that this "trick" would be a *very* good idea. We adopted it without further ado.)

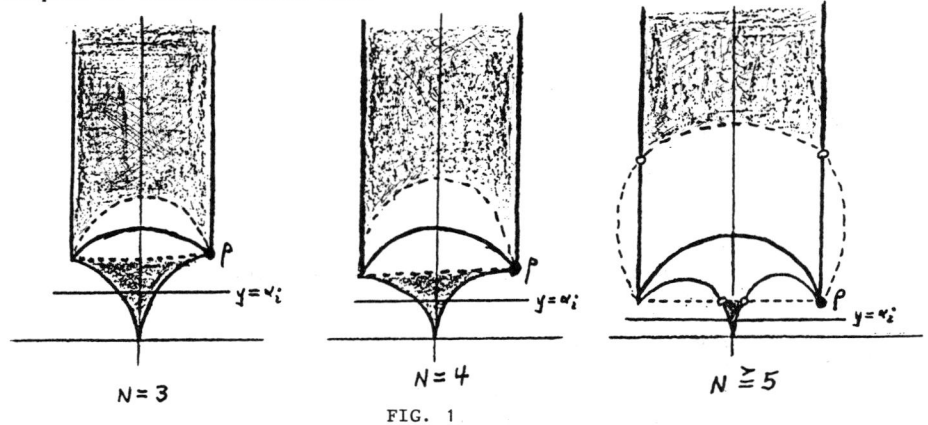

FIG. 1

Any number of *other* configurations were (actually) tested before we finally settled on type $(\alpha \| \beta)$. One curious finding was that the *vertical* batches used in [14] do *not* seem to condition so well once N starts to increase.

Our production jobs were all of type $(\alpha_1, \alpha_2 \| \beta_1, \beta_2)$. The parameters were as follows:

N = 4	N = 5	N = 6	N = 7
$(.60,.65 \| .62,.67)$	$(.45,.50 \| .47,.52)$	$(.40,.45 \| .42,.47)$	$(.30,.35 \| .32,.37)$
$(.50,.55 \| .52,.57)$	$(.40,.45 \| .42,.47)$	$(.35,.40 \| .37,.42)$	$(.22,.27 \| .24,.29)$
$\sin\frac{\pi}{4} = .70711$	$\sin\frac{\pi}{5} = .58779$	$\sin\frac{\pi}{6} = .50000$	$\sin\frac{\pi}{7} = .43388$
H1 = .025 ; H2 = .001 ; H3 = 10^{-6}			

As noted earlier, the algorithm outlined in §2 was implemented in standard CRAY-Fortran. In doing so: we were especially careful to arrange things so that, by deleting several z_j & w_j , it would ([1]) be possible to treat *several* M-values in parallel -- at least up to those points where (2.5) actually needed to be solved. ⟦This is done by appropriately structuring the "flow pattern" through levels $H1 >$ $H2 > H3$. ⟧

For safety: we (then) worked with *3* such M-values in our actual runs. Since the number of distinct $(\alpha \| \beta)$ types is *2* , this gives an effective total of 6 "tracks."

The choice of M changes with R . This is necessary to ensure "admissibility" in the sense of [14,eq.(2.6)] . That is: we need to have

$$(5.1) \qquad |I_\ell(z_j, R)| \; \lessapprox \; \left(\begin{array}{c} \text{something like} \\ 10^{-9} \end{array} \right) \cdot \max_{\substack{1 \le k \le M-1 \\ 1 \le n \le M}} |I_n(z_k, R)|$$

for $\ell > M$.

It is *not* wise to overshoot by too much on this aspect of the code.

The variability in M means that the grid points z_j & w_j must also be (occasionally) changed as well.

Fortunately: these changes are all very gradual.

For *each* $(\alpha \| \beta)$ type, we then prepared a list of semifinal R-values by scanning the 3 outputs (wrt M) for the best $|c_k' - c_k''|$ values.

To obtain the final R-values, we repeated this procedure (on the *2* semi-final lists).

The difference between the semifinal and final R-values was usually less than $\frac{1}{2} \times 10^{-6}$ and never greater than 1×10^{-6} .

Incidentally: declaring H2 to be *.001* ⟦even for smaller values of R⟧ is perfectly consistent with our remark in §3 about making provisional reductions. This particular choice seemed to work out fine...

The danger zone for nearly multiple eigenvalues will *therefore* be something like $|R_{n+1} - R_n| \lessapprox .005$.

Our final R_n-values are shown in Tables 1-4.

([1]) (in effect)

N = 4			
7.220872	16.138073*	20.530160	24.028513
9.533695*	16.644259*	21.049526	24.419715*
11.317680	17.493113	21.479057*	25.050855*
12.173008*	18.180918*	22.194674*	25.119336
13.310164	18.437078	22.374933	. . .
14.358510*	19.484714*	23.201396*	
15.274023	20.106695*	23.263712*	
Odd Eigenvalues for $\mathbb{C}(2\cos\frac{\pi}{4})$			
* indicates an oldform			

TABLE   1

N = 5			
6.473700	15.176893	19.385430	23.052526
8.636765	15.759928	19.962241	23.438611
10.136450	16.276410	20.597938	23.509476
11.015570	16.890976	20.745577	24.001860
12.084067	17.757303	21.287052	24.239718
12.851289	18.031441	21.675649	24.631401
14.071834	18.633434	22.197638	25.081315
14.307857	19.011695	22.399384	. . .
Odd Eigenvalues for $\mathbb{C}(2\cos\frac{\pi}{5})$			

TABLE   2

N = 6			
6.120576	15.483162	20.106695*	23.460177
8.193036	16.138073*	20.409439	24.209622
9.533695*	16.644259*	21.108696	24.419715*
10.507607	16.965398	21.479057*	24.916657
11.365904	17.820675	21.612650	24.952648
12.173008*	18.018977	22.100313	25.050855*
13.378621	18.180918*	22.194674*	...
13.507911	19.026777	22.671118	
14.358510*	19.484714*	23.201396*	
14.787325	19.566910	23.263712*	

Odd Eigenvalues for $\mathbb{C}(2\cos\frac{\pi}{6})$

* indicates an oldform

TABLE 3

N = 7			
5.921981	15.076620	19.463716	22.774514 ⊞
7.933889	15.625823	19.770114	23.186779 ⊞
9.185710	15.867575	20.374892	23.584303
10.229167	16.493571	20.655361	23.902652
10.905545	17.133710	20.919774	24.046762
11.803632	17.471380	21.331144	24.418591
12.851657	17.628681	21.449508	24.721778 ⊞
13.153879	18.324255	21.946932	24.884038
13.823102	18.767567	22.104986	25.220272
14.240129	18.868924	22.753843 ⊞	...

Odd Eigenvalues for $\mathbb{C}(2\cos\frac{\pi}{7})$

⊞ indicates a "troublesome" value; see §9(D) for details

23.186779 & 24.721778 seem to represent the *first* examples of §3 footnote 1

TABLE 4

Tables 5 & 6 supply some additional data. ⟨See also page 105 note 2.⟩
For information about CPU times, see §9(E).

N	type	$R \approx 10$	$R \approx 18$	$R \approx 25$
4	(.60,.65 ‖ .62,.67)	12	16	19
4	(.50,.55 ‖ .52,.57)	15	18	22
5	(.45,.50 ‖ .47,.52)	18	23	27
5	(.40,.45 ‖ .42,.47)	21	25	30
6	(.40,.45 ‖ .42,.47)	22	27	32
6	(.35,.40 ‖ .37,.42)	25	31	36
7	(.30,.35 ‖ .32,.37)	33	39	46
7	(.22,.27 ‖ .24,.29)	48	59	65
Sample M - values				

TABLE 5

N	R - range	2	3	4	5	6	7	8	9	10	12	14	16
4	10 ∿ 15	E-6	E-6	E-6	E-6	E-6	E-6	E-5	E-5	E-4	E-2	*	*
4	20 ∿ 25	E-6	E-6	E-6	E-6	E-6	E-6	E-6	E-6	E-5	E-4	E-3	E-2
5	10 ∿ 15	E-6	E-6	E-6	E-6	E-6	E-6	E-6	E-5	E-5	E-3	E-2	*
5	20 ∿ 25	E-6	E-6	E-6	E-6	E-6	E-6	E-6	E-6	E-6	E-6	E-5	E-5
6	10 ∿ 15	E-6	E-6	E-6	E-6	E-6	E-6	E-6	E-5	E-5	E-4	E-3	E-2
6	20 ∿ 25	E-6	E-6	E-6	E-6	E-6	E-6	E-6	E-6	E-6	E-6	E-5	E-5
7	10 ∿ 15	E-6	E-6	E-6	E-6	E-6	E-6	E-6	E-6	E-6	E-6	E-6	E-6
7	20 ∿ 25	E-6	E-6	E-6	E-6	E-6	E-6	E-6	E-6	E-6	E-6	E-6	E-6
Typical orders-of-magnitude for $\|c_k' - c_k''\|$													
using semifinal R_n - values													
N.B. The *best* cases are typically better by 1 or 2 orders.													

TABLE 6

Thus far we have emphasized R_n and *not* the associated Fourier coefficients.
Though the latter are *certainly* of interest, there seems to be very little point
in making complete lists of all the Fourier coefficients that were actually obtained.

In §7, we'll discuss 17 typical (*or* otherwise interesting!!) examples in
greater detail. See also §9(D).

Readers needing more information than this are advised to contact the author; the complete mass of Fourier coefficients is available on magnetic tape.

In scanning this output, we discovered *no* counterexamples to the Ramanujan-Petersson conjecture (for $N = 4,6$). Cf. [22] and equations (4.3), (4.11).

Before closing this section, we remark that we needed to make several additional test-runs for $N = 5$ and 7 in various ranges. These were necessary to help "calibrate" the corresponding output for the case of *even* R.

For the sake of completeness, we include a brief summary in Tables 7 & 8.

Note that several of the entries in table 7 make their way into §7.

N = 5 test run

30.029497
30.216067
30.578889
30.801263
30.859803
31.326651
31.429911
31.896429
32.161745
$29.875 \leq R \leq 32.375$
type $(.40,.45 \parallel .42,.47)$
$M = 37,39$
H1, H2 , H3 as before

TABLE 7a

N = 5 test run

50.020330
50.127645
50.308388
50.488237
50.537108
50.788214
50.907684
$49.875 \leq R \leq 51.000$
type $(.40,.45 \parallel .42,.47)$
$M = 48,50$
H1, H2 , H3 as before

TABLE 7b

N = 5 test run

79.965505
80.026912
80.129678
80.194123
80.497242
80.564910
80.618118
80.737485
80.838014
80.941330
$79.875 \leq R \leq 81.000$
type $(.45,.50 \parallel .47,.52)$
$M = 65,67$
H1, H2, H3 as before

TABLE   7c

N = 7 test run

50.196218
50.298439
50.446286
50.499022
50.663316
50.838387
50.891623
51.141579
$50.125 \leq R \leq 51.250$
type $(.30,.35 \parallel .32,.37)$
$M = 75,76,77$
H1, H2, H3 as before

TABLE 7d

N	R-range	2	3	4	5	6	7	8	9	10	12	14	16
5	≈ 30	E-5	E-5	E-5	E-5	E-5	E-5	E-5	E-5	E-5	E-5	E-5	E-5
5	≈ 50	E-6	E-6	E-6	E-6	E-6	E-6	E-6	E-6	E-6	E-6	E-6	E-5
5	≈ 80	E-5	E-5	E-5	E-5	E-5	E-5	E-5	E-5	E-5	E-5	E-5	E-5
7	≈ 50	E-6	E-6	E-6	E-6	E-6	E-6	E-6	E-6	E-6	E-6	E-6	E-6
Typical orders-of-magnitude for $\left|c_k' - c_k''\right|$													
using final R_n - values													
The "best" cases are typically better by 1 or 2 orders.													

TABLE 8

§6. <u>"Even" Eigenvalues for N = 4,5,6,7.</u> (*Compare* [45].) The situation for the arithmetic cases N = 4 and 6 is very similar to §5. Our *aim* was to compute the even eigenvalues 〚of $\mathfrak{C}_4$ and $\mathfrak{C}_6$〛 with $R \stackrel{<}{=} 25$ to an R - accuracy of six decimal places.

Our production jobs were all of type $(\alpha_1, \alpha_2 \| \beta_1, \beta_2)$. The parameters were as follows:

N = 4	N = 6
(.50,.55 ‖ .52,.57)	(.40,.45 ‖ .42,.47)
(.40..45 ‖ .42,.47)	(.35,.40 ‖ .37,.42)
$\sin\frac{\pi}{4} = .70711$	$\sin\frac{\pi}{6} = .50000$
H1 = .025 ; H2 = .001 ; H3 = 10^{-6}	

In an attempt to gain better accuracy, we decided to run things using *six* M - values (instead of just 3).

Our final R_n - values ⟦ for N = 4,6 ⟧ are displayed in Tables 9 & 10.

N = 4		
8.922877	17.878003	22.785909*
10.920392	19.125423	23.496586
13.779751*	19.423481*	24.112353*
14.685016	20.547604	24.856199
16.404109	21.315796*	25.052424
17.738563*	22.089045	...

Even Eigenvalues for $\mathbb{C}(2\cos\frac{\pi}{4})$

*   indicates an oldform

TABLE   9

N = 6		
5.098742	16.736215	21.807127
8.038861	17.500557	22.659272
9.743749	17.738563*	22.785908*
11.346418	18.647430	22.839291
11.889976	18.962642	23.620927
13.135144	19.423482*	23.979851
13.779751*	19.896104	24.112353*
14.626236	20.664907	24.293256
15.799494	21.315796*	24.931087
16.270959	21.434643	...

Even Eigenvalues for $\mathbb{C}(2\cos\frac{\pi}{6})$

*   indicates an oldform

TABLE   10

The discrepancy between the semifinal and final R - values was, except for the last 4 entries in Table 10, entirely similar to §5. ⟦The exceptions seemed to be caused by a conditioning problem with one of the types. The *other* type worked perfectly fine... ⟧

Tables 11 and 12(top) supply some additional data.

See §9(E) for information about CPU times and §7 for various examples illustrating the actual Fourier coefficients.

⟦No counterexamples to Ramanujan-Petersson were found in a scan through the total mass of computed c_k ... ⟧

— — —

The *focus* in the *non*arithmetic cases was (of course) quite different.

Here one basically wished to investigate conjecture (✯). We attacked this problem in the following ranges:

$$\left\{ \begin{array}{l} 0 \leqq R \leqq 60 \quad \text{for} \quad N = 5 \\ \hline 0 \leqq R \leqq 50 \quad \text{for} \quad N = 7 \end{array} \right\} \qquad \bullet$$

In both cases: *no even cusp forms were found.*

This assertion requires some elaboration, however.

Up to a point: the basic procedure is exactly like before. Our production jobs had the following parameters

N = 5	N = 7
(.50,.55 ‖ .52,.57) (.40,.45 ‖ .42,.47) (.30,.35 ‖ .32,.37) for R < 15	(.30,.35 ‖ .32,.37) (.22,.27 ‖ .24,.29) for R < 40
(.45,.50 ‖ .47,.52) (.40,.45 ‖ .42,.47) for R > 15	(.22,.27 ‖ .24,.29) (.19,.24 ‖ .21,.26) for R > 40
$\sin\frac{\pi}{5} = .58779$	$\sin\frac{\pi}{7} = .43388$
H1 = .025 ; H2 = .001 ; H3 = 10^{-6}	

$\bullet$

For N = 5 and $0 \leqq R \leqq 15$, we ran *6* (and sometimes 9) M-values in parallel. In all other cases: we used 3 .

In §3, we explained why it is important to keep the numbers α_i' & β_k below $\sin(\frac{\pi}{N})$. Placing these levels *too low*, however, causes M to become rather large — which begins to affect the overall accuracy (and CPU time) adversely. It is *therefore* necessary to strike some kind of balance.

One *might* think that running jobs with the parameters shown above would simply produce no output ⟦in accordance with (✗)⟧ . This, however, is *not* the case.

The assertion that "no even cusp forms were found" is not as simple as it looks.

What typically happened in our (even) nonarithmetic jobs was that R–values would occasionally come out showing differences $\left|c_k' - c_k''\right|$ that looked "half-way" respectable.

BUT: *one or more* danger signs would *invariably* apply.

These signs included:

(a) excessive movement (or disappearance!) of the proposed R–value when M is varied;

(b) excessive movement (or disappearance!) of the proposed R–value when the "type" is varied;

(c) excessive movement in the first few Fourier coefficients under similar variations;

(d) values of $\left|c_k' - c_k''\right|$ that were typically 3 to 4 orders-of-magnitude *worse* than their "counterparts" for *odd* R (in the same range).

Item (d), on its own, was usually enough to destroy any putative R .

The essential point here is that one should expect similar levels of "stability" to be exhibited by both the even & odd R–values. This was certainly the case for N = 4 , 6 . (^{1})

One always has to be a bit careful in situations like this to exclude the possibility that some type of intrinsic "static" region (wrt R) is the real culprit in (a)-(d). On such R–regions, there could easily be an overall degradation in conditioning-level ⟦ which causes (a)-(d)⟧ .

This effect was discussed in [14] at some length. The *same* effect certainly occurs for N = 4,5,6,7 . In fact: here it begins even earlier ⟦with "missing" R_n–values occurring on one-or-another track for R as low as 9.533 ⟧ .

At this stage of the game, one can only *hope* that the (potential) effects of such "static" are *indeed* minimized by running several tracks and/or types. (^{2})

(^{1}) Though this expectation seems reasonable enough for arbitrary N , we make *no* pretenses about having a rigorous proof. The full statement for N = 5,7 should therefore read: no even cusp forms *meeting a reasonable set of standards* were detected by the machine. ⟦Matters are further complicated by §3 footnote 1 ... ⟧

(^{2}) (the rationale being that such "static" regions should occur *randomly* and with *small* relative measure)

In this connection: the most interesting thing is that some R-values with half-way decent $|c_k' - c_k''|$ *did* manage to stay reasonably "intact" ⟦ wrt (a)-(c) ⟧ under *several* changes of track.

For both N = 5 and 7 , such R-values seemed to be most common ⟦*and* "strongest"⟧ in cases where α & β could come closest to $\sin(\frac{\pi}{N})$.

In view of this $\alpha\beta$ - dependence, it is natural to conjecture that the foregoing R-values correspond to *some kind* of "residual" effect from the pseudo cusp forms mentioned in §3 ⟦*as opposed* to something like (1.2)⟧ .

This interpretation does, in fact, seem to be correct.

If one repeatedly tests various hybrid (i.e. partially *un*safe) batches with (say) N = 5 , the necessary patterns begin to emerge very clearly. For further details on this, see §8.

⟦An *alternate* way of reaching the same conclusion is to study the effect of the 0-order term in (1.2) on the linear algebra in (2.5). To make $|c_n' - c_n''|$ reasonably small for the first few n , one would "basically" need $\sqrt{\eta}$ to be something like 10^{-2} (or less). It seems rather *far-fetched* to believe that $E(z;s;\mathfrak{C}_N)$ is going to have (a dozen or more) poles situated within 10^{-4} units of { Re(s) = $\frac{1}{2}$ } when N is only 5 and $|Im(s)|$ is no more than 25... ⟧

The upshot of these remarks is very simple. The occurrence of "pseudo-residuals" makes it *doubly* important to pay close attention to (a)-(d) ⟦*and to* the size of α_i' & β_k ⟧. Failure to do so may cause one to "snare" the wrong type of "animal" altogether...

Tables 11 and 12 provide some additional information about our "even" runs. Compare: tables 6,8,5 in §5.

N	R-range	2	3	4	5	6	7	8	9	10	12	14	16
4	10 ∼ 15	E-7	E-7	E-7	E-6	E-6	E-6	E-6	E-6	E-4	E-2	*	*
4	20 ∼ 25	E-6	E-6	E-6	E-6	E-6	E-6	E-5	E-5	E-5	E-5	E-4	E-2
6	10 ∼ 15	E-7	E-7	E-7	E-7	E-6	E-6	E-6	E-5	E-5	E-3	E-1	*
6	20 ∼ 25	E-7	E-7	E-7	E-7	E-7	E-6	E-6	E-6	E-6	E-5	E-5	E-5
Typical orders-of-magnitude for $|c_k' - c_k''|$													
using semifinal R_n-values													
N.B. The *best* cases are typically better by 1 or 2 orders.													

TABLE   11

N	type	R $\approx$ 10	R $\approx$ 18	R $\approx$ 25	R $\approx$ 40	R $\approx$ 50	R $\approx$ 60
4	(.50,.55 ‖ .52,.57)	15	18	22	*	*	*
4	(.40,.45 ‖ .42,.47)	17	21	25	*	*	*
6	(.40,.45 ‖ .42,.47)	21	26	31	*	*	*
6	(.35,.40 ‖ .37,.42)	23	29	34	*	*	*
5	(.45,.50 ‖ .47,.52)	*	23	26	40	46	52
5	(.40,.45 ‖ .42,.47)	21	25	29	44	51	58
7	(.30,.35 ‖ .32,.37)	34	40	48	64	*	*
7	(.22,.27 ‖ .24,.29)	48	59	72	96	106	*
7	(.19,.24 ‖ .21,.26)	*	*	*	110	122	*

Sample M - values

TABLE 12

§7. <u>Some Examples</u>. In this section, we'll look at 17 examples which serve to illustrate various aspects of our production runs. The information given in each case will include: appropriately rounded values of $\frac{1}{2}(c_k' + c_k'')$; a rough indication of $|c_k' - c_k''|$; and a brief description of the "track" used.

When discussing old-forms, remember that:

$$C_n = a_n + \sqrt{q}\, \frac{a_n}{q}$$

by virtue of (4.6).

Example 1.   R = 7.220872   (N = 4 / odd / new-form) .

k	$\frac{1}{2}(c_k' + c_k'')$	rough diff.	k	$\frac{1}{2}(c_k' + c_k'')$	rough diff.
2	−.7071067	2E−8	7	−.0625	5E−4
3	−.9493510	3E−8	8	−.3460	4E−3
4	.5000021	5E−7	9	−.142	3E−2
5	−.869730	6E−6	10	.80	0.1
6	.671435	6E−5	11	−.57	0.3
type (.60,.65 ‖ .62,.67)		M = 11		final R	

As an indication of (overall) accuracy, note that:

$$\frac{1}{\sqrt{2}} = .707106781 \quad , \quad \left|c_2 + \frac{1}{\sqrt{2}}\right| = .0000001 \quad , \quad \left|c_4 - \frac{1}{2}\right| = .0000021$$

$$\left|c_6 - c_2 c_3\right| = .000143 \qquad \left|c_8 + \frac{1}{2\sqrt{2}}\right| = .0076$$

$$\left|c_9 - (c_3^2 - 1)\right| = .043 \qquad \left|c_{10} - c_2 c_5\right| = .19 \qquad \bullet$$

<u>Example 2</u>.   R = 12.173008   (N = 4 / odd / old-form) .

k	$\frac{1}{2}(c_k' + c_k'')$	rough diff.	a_k
2	1.7034654	1E-8	.2892518
3	-1.2018588	1E-8	-1.2018588
4	-.5072694	4E-9	-.9163332
5	.0395527	6E-11	.0395527
6	-2.0473248	2E-9	-.3476398
7	.4481331	3E-8	.4481331
8	-1.8501922	3E-7	-.5543014
9	.4444580	6E-7	.4444580
10	.06754	6E-5	.01160
11	-.6935	6E-4	-.6935
12	.638	1E-2	1.130
13	-3.11	1.25(!!)	-3.11
type (.50,.55 ‖ .52,.57) M = 17			final R

As an indication of (overall) accuracy, note that:

$$\left|a_4 - (a_2^2 - 1)\right| = .0000002 \qquad \left|a_6 - a_2 a_3\right| = .0000000$$

$$\left|a_8 - (a_2^3 - 2a_2)\right| = .0000015 \qquad \left|a_9 - (a_3^2 - 1)\right| = .0000066$$

$$\left|a_{10} - a_2 a_5\right| = .00016 \qquad \left|a_{12} - a_3 a_4\right| = .029 \qquad \bullet$$

In view of the (large) difference at k = 13, we do *not* take c_{13} seriously.

The situation for $k \gtrless 14$ gets progressively worse. In the terminology of [14] , we can thus say that the c_n "hump" occurs at about 13 $\sim$ 14 .

Incidentally: observe that

$$a_2 = \left\{\begin{array}{ll} .2892518 & \text{here} \\ .289252 & \text{in [14]} \end{array}\right\} \qquad a_3 = \left\{\begin{array}{ll} -1.2018588 & \text{here} \\ -1.201858 & \text{in [14]} \end{array}\right\}$$

$$a_5 = \left\{\begin{array}{ll} .0395527 & \text{here} \\ .042 & \text{in [14]} \end{array}\right\} \qquad a_7 = \left\{\begin{array}{ll} .4481331 & \text{here} \\ *** & \text{in [14]} \end{array}\right\} \qquad \bullet$$

The current a_k - listing is (thus) significantly better than the one in [14] .

This improvement basically reflects the *change* in geometry.

Example 3.   R = 24.028513   (N=4 / odd / new-form) .

k	$\frac{1}{2}(c_k' + c_k'')$	rough diff.	k	$\frac{1}{2}(c_k' + c_k'')$	rough diff.
2	−.7071066	3E-8	11	−.879083	1E-6
3	.5772141	2E-8	12	.28860	5E-5
4	.5000000	4E-8	13	1.6529	2E-4
5	.2392995	7E-8	14	.0854	3E-4
6	−.4081516	5E-8	15	.143	3E-3
7	−.1212567	2E-7	16	.247	8E-4
8	−.3535511	3E-7	17	−1.09	0.17
9	−.666826	2E-6	18	.68	0.45
10	−.169205	4E-6			
type (.50,.55 ‖ .52,.57) M = 22				final R	

To indicate the overall accuracy, note that:

$$\left| c_2 + \frac{1}{\sqrt{2}} \right| = .0000002 \qquad \left| c_4 - \frac{1}{2} \right| = .0000000$$

$$\left| c_6 - c_2 c_3 \right| = .0000003 \qquad \left| c_8 + \frac{1}{2\sqrt{2}} \right| = .0000023$$

$$\left| c_9 - (c_3^2 - 1) \right| = .000002 \qquad \left| c_{10} - c_2 c_5 \right| = .000005$$

$$\left| c_{12} - c_3 c_4 \right| = .000007 \qquad \left| c_{14} - c_2 c_7 \right| = .0003$$

$$\left| c_{15} - c_3 c_5 \right| = .005 \qquad \left| c_{16} - \frac{1}{4} \right| = .003 \qquad \bullet$$

The c_n "hump" occurs at about n = 18.

— — — — —

Before moving onward, we need to draw attention to an important fact. By reviewing examples 1-3, it becomes apparent that the number $\left| c_k' - c_k'' \right|$ is *not* a true indicator of the actual *error* in c_k . (This is seen by looking at the multiplicative relations.) To be on the safe side, it seems preferable to use something like $\max\left[2 \times 10^{-7}, 5(\text{diff}) \right]$ as "the basic indicator" of fuzz-level. [1]

[1] To *mollify* purely random effects, (diff) should actually be replaced here by some type of backward average. [Unless $\left| c_k' - c_k'' \right|$ is abnormally small compared to its neighbors, this modification is usually insignificant...]

Example 4.   R = 6.120576   (N=6 / odd / new-form) .

k	$\frac{1}{2}(c_k' + c_k'')$	rough diff.	k	$\frac{1}{2}(c_k' + c_k'')$	rough diff.
2	−.6716156	1E−9	10	−.21872	3E−5
3	−.5773503	4E−10	11	−.68593	3E−5
4	−.5489325	4E−9	12	.3176	8E−4
5	.3256987	2E−9	13	−.3325	8E−4
6	.3877575	6E−8	14	.99	2E−2
7	−1.4557169	4E−8	15	−.17	2E−2
8	1.040288	1E−6	16	.30	0.64
9	.333334	1E−6			
type (.40,.45 ‖ .42,.47) M = 20				final R	

To indicate overall accuracy, note that:

$$\frac{1}{\sqrt{3}} = .577350269 \qquad\qquad \left|c_3 + \frac{1}{\sqrt{3}}\right| = .0000000$$

$$\left|c_4 - (c_2^2 - 1)\right| = .0000000 \qquad\qquad \left|c_6 - c_2 c_3\right| = .0000000$$

$$\left|c_8 - (c_2^3 - 2c_2)\right| = .0000008 \qquad\qquad \left|c_9 - \frac{1}{3}\right| = .000001$$

$$\left|c_{10} - c_2 c_5\right| = .00002 \qquad\qquad \left|c_{12} - c_3 c_4\right| = .0007$$

$$\left|c_{14} - c_2 c_7\right| = .012 \qquad\qquad \left|c_{15} - c_3 c_5\right| = .02 \qquad\qquad .$$

The c_n "hump" occurs at about n = 16.

Example 5.   R = 12.173008   (N=6 / odd / old-form) .

k	$\frac{1}{2}(c_k' + c_k'')$	rough diff.	a_k
2	.2892519	6E−8	.2892519
3	.5301920	2E−8	−1.2018588
4	−.9163334	3E−8	−.9163334
5	.0395526	6E−8	.0395526
6	.1533590	8E−8	−.3476400
7	.4481331	8E−8	.4481331
8	−.5543028	2E−7	−.5543028
9	−1.6372162	4E−7	.4444643
10	.011442	2E−6	.011442
11	−.691455	7E−6	−.691455
12	−.48581	3E−5	1.10133
13	−.8030	3E−4	−.8030
14	.132	3E−3	.132

15	.02	8E-3	-.05
16	.72	4E-2	.72
17	-.7	0.41	-.7
type $(.35,.40 \| .37,.42)$ M = 27			a semifinal R

To indicate the overall accuracy, note that:

$$|a_4 - (a_2^2 - 1)| = .0000001 \qquad\qquad |a_6 - a_2 a_3| = .0000001$$

$$|a_8 - (a_2^3 - 2a_2)| = .0000003 \qquad |a_9 - (a_3^2 - 1)| = .0000003$$

$$|a_{10} - a_2 a_5| = .000001 \qquad\qquad |a_{12} - a_3 a_4| = .000027$$

$$|a_{14} - a_2 a_7| = .002 \qquad\qquad |a_{15} - a_3 a_5| = .002$$

$$|a_{16} - (a_2^4 - 3a_2^2 + 1)| = .036$$

The c_n "hump" occurs at about n = 18.

The current a_k - values are an improvement over those in example 2.

It is also interesting to compare things with the old result in [13,p.653], [19] :

R = 12.1730083246797

a_2 = .2892518714	a_5 = .03955272
a_3 = -1.201858761	a_7 = .4481

These numbers were obtained using double-precision arithmetic (and an *un*inspired Newton-Cotes type algorithm for the K-Bessel function).

Example 6.   R = 24.419715   (N=6 / odd / old-form) .

k	$\frac{1}{2}(c_k' + c_k'')$	rough diff.	a_k
2	.9655410	2E-7	.9655410
3	1.0417911	2E-7	-.6902597
4	-.0677319	2E-7	-.0677319
5	1.3158034	2E-7	1.3158034
6	1.0058915	5E-7	-.6664746
7	-.5454961	6E-7	-.5454961
8	-1.0309378	2E-8	-1.0309378
9	-1.719106	2E-6	-.523541
10	1.270463	2E-6	1.270463
11	-.156968	3E-6	-.156968
12	-.070563	3E-6	.046752
13	-1.894287	7E-6	-1.894287
14	-.526697	4E-6	-.526697
15	1.370791	4E-6	-.908247

16	-.927680	5E-6	-.927680
17	.344743	8E-6	.344743
18	-1.659862	1E-5	-.505494
19	-.10064	2E-5	-.10064
20	-.08913	2E-5	-.08913
21	-.56832	6E-5	.37651
22	-.15148	2E-4	-.15148
23	-.70406	4E-5	-.70406
24	-1.07402	7E-6	.71162
25	.7321	2E-3	.7321
26	-1.8307	3E-3	-1.8307
27	.156	2E-2	1.063
28	.028	2E-2	.028
29	.360	4E-2	.360
30	1.27	1E-1	-.93
31	.29	4E-2	.29
32	-.09	4E-1	-.09
type (.35,.40 ‖ .37,.42) M = 37			a typical final R (*with* somewhat larger M)

As an indication of (overall) accuracy, note that:

$$|a_4 - (a_2^2-1)| = .0000013 \qquad |a_6 - a_2 a_3| = .0000006$$

$$|a_8 - (a_2^3-2a_2)| = .0000002 \qquad |a_9-(a_3^2-1)| = .0000005$$

$$|a_{10} - a_2 a_5| = .000001 \qquad |a_{12} - a_3 a_4| = .000001$$

$$|a_{14} - a_2 a_7| = .000002 \qquad |a_{15} - a_3 a_5| = .000001$$

$$|a_{16}-(a_2^4-3a_2^2+1)| = .000002 \qquad |a_{18} - a_2 a_9| = .000006$$

$$|a_{20} - a_4 a_5| = .000008 \qquad |a_{21} - a_3 a_7| = .00002$$

$$|a_{22} - a_2 a_{11}| = .00008 \qquad |a_{24} - a_3 a_8| = .00001$$

$$|a_{25}-(a_5^2 - 1)| = .0008 \qquad |a_{26} - a_2 a_{13}| = .0017$$

$$|a_{27} - (a_3^3-2a_3)| = .011 \qquad |a_{28} - a_4 a_7| = .009$$

$$|a_{30} - a_2 a_{15}| = .053 \qquad |a_{32} - (a_2^5-4a_2^3+3a_2)| = .23 \qquad \bullet$$

The c_n "hump" occurs at about $32 \sim 33$.

By way of comparison to [14] , observe that:

$$a_2 = \left\{ \begin{array}{ll} .9655410 & \text{here} \\ .965541 & \text{in } [14] \end{array} \right\} \qquad a_3 = \left\{ \begin{array}{l} -.6902597 \\ -.690260 \end{array} \right\}$$

$$a_5 = \left\{ \begin{array}{ll} 1.3158034 & \text{here} \\ 1.315804 & \text{in } [14] \end{array} \right\} \qquad a_7 = \left\{ \begin{array}{l} -.5454961 \\ -.545 \end{array} \right\} \qquad \bullet$$

* * * * * *

<u>Example 7.</u>   R = 13.77975137   (N=4 / even / old-form) .

k	$\frac{1}{2}(c_k' + c_k'')$	rough diff.	a_k
2	2.9635181	3E-9	1.5493045
3	.2468996	3E-9	.2468996
4	3.5913914	9E-8	1.4003440
5	.7370610	9E-7	.7370610
6	.7316926	3E-7	.3825238
7	-.2614212	8E-7	-.2614212
8	2.600645	2E-6	.620260
9	-.939045	2E-6	-.939045
10	2.1825	2E-3	1.1401
11	-.960	7E-3	-.960
12	1.15	0.37	.61
type (.40,.45 ‖ .42,.47) M = 19			final R

For the overall accuracy, note that:

$$|a_4 - (a_2^2-1)| = .0000004 \qquad |a_6 - a_2 a_3| = .0000011$$

$$|a_8 - (a_2^3-2a_2)| = .000005 \qquad |a_9 - (a_3^2-1)| = .000004$$

$$|a_{10} - a_2 a_5| = .0018 \qquad |a_{12} - a_3 a_4| = .26 \qquad \bullet$$

The c_n "hump" occurs at about n = 12.

To make a comparison with [14,41] , observe that:

$$R = \left\{ \begin{array}{ll} 13.77975137 & \text{here} \\ 13.77975135189 & \text{in } [14,\S 10] \\ 13.7797513519 & \text{in } [41] \end{array} \right\}$$

$$a_2 = \left\{ \begin{array}{ll} 1.5493045 & \text{here} \\ 1.54930447794 & \text{in } [14,\S 10] \\ 1.5493044779 & \text{in } [41] \end{array} \right\} \qquad a_3 = \left\{ \begin{array}{l} .2468996 \\ .24689977245 \\ .2468997725 \end{array} \right\}$$

$$a_5 = \left\{ \begin{array}{l} .7370610 \\ .737060383 \\ .7370603853 \end{array} \right\} \qquad a_7 = \left\{ \begin{array}{l} -.2614212 \\ -.261421 \\ -.2614200758 \end{array} \right\}$$

$$a_{11} = \left\{ \begin{array}{c} -.960 \\ *** \\ -.9535646526 \end{array} \right\} \qquad \bullet$$

Our earlier remark about c_k-error is nicely illustrated at $k = 3$. [Errors of this kind appear to stem mainly from the fact that we chose H3 to be 10^{-6} in all our production runs. Cf. §§2,5,6. A reduction in H3 should yield better accuracy...]

For the sake of completeness, we also recall that [13,p.653], [19] had:

R = 13.7797513518907

a_2 = 1.54930447794	a_3 = .24689977245
a_5 = .7370604	a_7 = -.2614

Example 8.   R = 17.878003   (N=4 / even / new-form) .

k	$\frac{1}{2}(c_k' + c_k'')$	rough diff.	k	$\frac{1}{2}(c_k' + c_k'')$	rough diff.
2	-.707106798	7E-11	10	-1.2811050	2E-6
3	.9825686	4E-12	11	-.1747944	5E-7
4	.5000000	2E-9	12	.491474	1E-6
5	1.8117563	1E-9	13	1.10189	7E-5
6	-.6947809	4E-8	14	-.0101	1E-3
7	.0082109	3E-8	15	1.795	6E-3
8	-.3535540	4E-7	16	.261	2E-3
9	-.0345570	1E-6	17	.43	0.14
type (.40,.45 ‖ .42,.47) M = 21				final R	

To indicate the overall accuracy, note that:

$$\left| c_2 + \frac{1}{\sqrt{2}} \right| = .000000017 \qquad \left| c_4 - \frac{1}{2} \right| = .0000000$$

$$\left| c_6 - c_2 c_3 \right| = .0000000 \qquad \left| c_8 + \frac{1}{2\sqrt{2}} \right| = .0000006$$

$$\left| c_9 - (c_3^2 - 1) \right| = .0000019 \qquad \left| c_{10} - c_2 c_5 \right| = .0000002$$

$$\left| c_{12} - c_3 c_4 \right| = .000190 \qquad \left| c_{14} - c_2 c_7 \right| = .0043$$

$$\left| c_{15} - c_3 c_5 \right| = .015 \qquad \left| c_{16} - \frac{1}{4} \right| = .01 \qquad \bullet$$

The c_n "hump" occurs at about $17 \sim 18$.

* * * * * *

Example 9.   R = 14.626236   (N=6 / even / new-form).

k	$\frac{1}{2}(c_k' + c_k'')$	rough diff.	k	$\frac{1}{2}(c_k' + c_k'')$	rough diff.
2	.55536214	7E-10	10	-.5291602	9E-7
3	-.57735035	3E-8	11	-.015748	4E-6
4	-.6915731	3E-7	12	.39927	2E-5
5	-.9528186	3E-7	13	1.0297	1E-3
6	-.3206385	3E-7	14	1.017	9E-3
7	1.8374419	3E-7	15	.66	1E-2
8	-.9394356	1E-7	16	-.05	0.17
9	.3333338	3E-7			

type (.40,.45 ‖.42,.47) M = 22	final R

As an indication of over-all accuracy, note that:

$$\left|c_3 + \frac{1}{\sqrt{3}}\right| = .00000008 \qquad \left|c_4 - (c_2^2 - 1)\right| = .0000002$$

$$\left|c_6 - c_2 c_3\right| = .0000000 \qquad \left|c_8 - (c_2^3 - 2c_2)\right| = .0000001$$

$$\left|c_9 - \frac{1}{3}\right| = .0000005 \qquad \left|c_{10} - c_2 c_5\right| = .0000008$$

$$\left|c_{12} - c_3 c_4\right| = .00001 \qquad \left|c_{14} - c_2 c_7\right| = .003$$

$$\left|c_{15} - c_3 c_5\right| = .11 \qquad \left|c_{16} - (c_2^4 - 3c_2^2 + 1)\right| = .22$$

The c_n "hump" occurs at about n = 16.

Example 10.   R = 13.7797513515   (N=6 / even / old-form) .

k	$\frac{1}{2}(c_k' + c_k'')$	rough diff.	a_k
2	1.549304477	5E-11	1.549304477
3	1.978950582	6E-10	.246899774
4	1.400344368	1E-9	1.400344368
5	.737060386	1E-8	.737060386
6	3.065997001	1E-8	.382522930
7	-.26142006	1E-8	-.26142006
8	.62025531	1E-8	.62025531
9	-.5113975	3E-8	-.9390405
10	1.1419309	9E-8	1.1419309
11	-.9535642	3E-7	-.9535642
12	2.771212	3E-6	.345744
13	.278822	2E-5	.278822
14	-.40516	2E-5	-.40516

15	1.464	4E-3	.187
16	-.46	2E-2	-.46
17	1.17	0.11	1.17

| type (.40,.45 ‖ .42,.47) M = 20 | one of our strongest final R |

To indicate the over-all accuracy, note that:

$$\left| a_4 - (a_2^2 - 1) \right| = .000000006 \qquad \left| a_6 - a_2 a_3 \right| = .000000005$$

$$\left| a_8 - (a_2^3 - 2a_2) \right| = .00000000 \qquad \left| a_9 - (a_3^2 - 1) \right| = .0000000$$

$$\left| a_{10} - a_2 a_5 \right| = .0000001 \qquad \left| a_{12} - a_3 a_4 \right| = .000001$$

$$\left| a_{14} - a_2 a_7 \right| = .00014 \qquad \left| a_{15} - a_3 a_5 \right| = .005$$

$$\left| a_{16} - (a_2^4 - 3a_2^2 + 1) \right| = .02 \qquad \qquad \bullet$$

The c_n "hump" occurs at about $17 \sim 18$.

It is also interesting to compare things with [14,13,19,41] as in example 7.

	here	[41] ; [13,p.729]	[14,§10]	[13,p.653]
R-13	.7797513515	.77975135189	.77975135189	.77975135189
a_2	1.549304477	1.54930447794	1.54930447794	1.54930447794
a_3	.246899774	.24689977245	.24689977245	.24689977245
a_5	.737060386	.73706038534	.737060383	.7370604
a_7	-.26142006	-.26142007577	-.261421	-.2614
a_{11}	-.9535642	-.95356465262	***	***
a_{13}	.278822	.27882702916	***	***
a_{17}	1.17	1.30734171453	***	***

The present accuracy is very striking because taking $H3 = 10^{-6}$ would ordinarily suggest $7 \sim 7\frac{1}{2}$ decimal places as being the upper limit. 〖Bear in mind *too* that M = 20 , and that we are using only single-precision arithmetic... 〗

Example 11.   R = 17.73856338   (N=6 / even / old-form) .

k	$\frac{1}{2}(c_k' + c_k'')$	rough diff.	a_k
2	-.76545805	2E-10	-.76545805
3	.75427190	3E-11	-.97777891
4	-.41407396	1E-9	-.41407396
5	-1.01527351	1E-11	-1.01527351
6	-.57736350	2E-9	.74844873
7	1.18082083	7E-9	1.18082083

8	1.08241430	5E-9	1.08241430
9	-1.7375111	4E-8	-.0439483
10	.7771493	2E-8	.7771493
11	-.6204877	1E-7	-.6204877
12	-.312325	3E-6	.404872
13	.265291	2E-5	.265291
14	-.90386	2E-5	-.90386
15	-.7659	5E-4	.9926
16	-.414	2E-3	-.414
17	-.135	2E-3	-.135
18	1.33	2E-2	.034
19	.18	.09	.18

type $(.40,.45 \| .42,.47)$ M=27	one of our strongest final R

For the overall accuracy, note that:

$$|a_4 - (a_2^2 - 1)| = .00000001 \qquad |a_6 - a_2 a_3| = .00000001$$

$$|a_8 - (a_2^3 - 2a_2)| = .00000001 \qquad |a_9 - (a_3^2 - 1)| = .0000001$$

$$|a_{10} - a_2 a_5| = .0000000 \qquad |a_{12} - a_3 a_4| = .000001$$

$$|a_{14} - a_2 a_7| = .00001 \qquad |a_{15} - a_3 a_5| = .0001$$

$$|a_{16} - (a_2^4 - 3a_2^2 + 1)| = .000 \qquad |a_{18} - a_2 a_9| = .000$$

The c_n "hump" occurs at about $n = 20$.

A comparison with [13,pp.653,729] gives:

	here	[13,p.729]	[13,p.653]
R	17.73856338	17.7385633811	17.7385633811
a_2	-.76545805	-.76545806	-.7654580566
a_3	-.97777891	-.9777789	-.9777789075
a_5	-1.01527351	-1.0152735	-1.0152735
a_7	1.18082083	1.1808208	1.1807
a_{11}	-.6204877	-.6204877	***
a_{13}	.265291	.2652887	***
a_{17}	-.135	-.1357404	***

The results in the middle column were obtained by H.Stark using the *same* method as in [41] .

The present accuracy is again rather striking.

* * * * * *

Example 12. R = 14.071834 (N=5 / odd) .   In a *nonarithmetic* case like N = 5 , it is not so clear what will happen. We therefore look at *both* semifinal values.

	R = 14.0718340			R = 14.0718335	
k	$\frac{1}{2}(c_k' + c_k'')$	rough diff.	k	$\frac{1}{2}(c_k' + c_k'')$	rough diff.
2	−.3658834	3E−9	2	−.365882	1E−6
3	−.5092615	1E−8	3	−.509263	2E−6
4	.5249263	3E−7	4	.524923	5E−6
5	.0132187	2E−7	5	.013217	2E−7
6	.169034	5E−7	6	.169027	4E−6
7	.509436	4E−8	7	.509436	5E−8
8	1.007591	2E−7	8	1.007585	6E−7
9	−.477043	2E−7	9	−.477040	1E−6
10	−1.113400	4E−7	10	−1.11347	3E−5
11	−.178021	2E−6	11	−.17791	7E−5
12	−1.038675	2E−6	12	−1.040	1E−3
13	−.53924	2E−5	13	−.536	2E−3
14	−1.2025	3E−4	14	−1.25	4E−2
15	.639	2E−3	15	.72	.08
16	−.04	.06	16	−1.5	1.56
left type (.40,.45 ‖.42,.47) M = 22			typical semifinal		
right type (.45,.50‖ .47,.52) M = 18			R − values		

The c_n "humps" occur at about 17 and 15 , respectively. The agreement between the 2 columns (of c_n) is consistent with our earlier remark about fuzz-level. In view of table 5, the right-hand differences are just about average. The ones on the *left* are significantly better. ⟦Though in this example there was some *advantage* to keeping α_i & β_k further away from $\sin(\frac{\pi}{5})$, the overall situation is basically random.⟧

There are no (obvious) multiplicative relations.

Example 13. R = 14.307857 (N=5 / odd). This example illustrates the possi-bility of *large* c_n occurring when $\mathfrak{C}_N$ is non-arithmetic.

	R = 14.3078567			R = 14.3078568	
k	$\frac{1}{2}(c_k' + c_k'')$	rough diff.	k	$\frac{1}{2}(c_k' + c_k'')$	rough diff.
2	10.37429	6E−5	2	10.37433	4E−4
3	−3.15814	2E−5	3	−3.15815	1E−4
4	6.954667	2E−6	4	6.95462	1E−4
5	−5.74760	1E−4	5	−5.74765	2E−4
6	−7.46890	3E−5	6	−7.4690	4E−4
7	−7.69329	8E−5	7	−7.6933	3E−4
8	3.48202	3E−5	8	3.4819	1E−4
9	−6.6204	1E−4	9	−6.6203	3E−6

10	2.5325	2E-4	10	2.531	3E-3
11	.2486	8E-4	11	.251	4E-3
12	-4.893	4E-3	12	-4.93	6E-2
13	-.19	9E-2	13	-.01	0.11
14	6.96	0.29	14	5.9	1.62

left type (.40,.45 ‖ .42,.47) M = 23	worse than average
right type (.45,.50 ‖ .47,.52) M = 18	semifinal R

The c_n "humps" occur at about 15 and 14, respectively. The agreement between the c_n's is consistent with our rule-of-thumb. Though the differences are below average in quality, things don't look so bad if one deals with *significant* figures instead. 〖On a floating-point machine, this might not be such a bad idea... 〗

Other cases having relatively large c_n are:

$$R = 25.081315 \quad (\text{e.g. } c_2 = -5.703319 \,,\quad c_3 = -9.051076) \quad ;$$

$$R = 30.029497 \quad (\text{e.g. } c_2 = -12.67644 \,,\quad c_3 = 22.34885) \quad .$$

For $N = 7$, the situation is even sharper:

$$R = 23.186779 \quad (\text{e.g. } c_2 = 108.484 \,,\quad c_3 = 117.505) \quad ;$$

$$R = 24.721778 \quad (\text{e.g. } c_2 = -102.540 \,,\quad c_3 = -46.273) \quad .$$

Example 14.   R = 50.488237 (N=5 / odd) .

k	$\frac{1}{2}(c_k' + c_k'')$	rough diff.	k	$\frac{1}{2}(c_k' + c_k'')$	rough diff.
	R = 50.48823748			R = 50.48823704	
2	-3.814250	8E-6	2	-3.81433	1E-4
3	3.965753	8E-6	3	3.96583	1E-4
4	1.162531	2E-6	4	1.16255	3E-5
5	1.821024	5E-6	5	1.82104	2E-5
6	-1.664537	1E-6	6	-1.66457	7E-5
7	1.012407	2E-6	7	1.01242	3E-5
8	.608971	6E-6	8	.608989	6E-7
9	-.157489	1E-6	9	-.157488	6E-6
10	-1.592329	6E-7	10	-1.592330	5E-6
12	-3.305690	5E-6	12	-3.30577	1E-4
14	2.548589	2E-6	14	2.54862	1E-4
16	3.509777	7E-6	16	3.50983	1E-4
18	2.861370	3E-5	18	2.86141	6E-5
20	-.036687	2E-6	20	-.03668	3E-5
25	4.55179	5E-5	25	4.55174	4E-4
30	-.41234	2E-5	30	-.41239	1E-4
35	3.9363	2E-4	35	3.93652	1E-5
40	.3673	5E-3	40	.3671	2E-3
45	-.18	1.11	45	1.14	0.61

type (.40,.45 ‖ .42,.47)	left M = 48	a fairly typical case
	right M = 50	illustrating variation wrt M

The purpose of this example is to show how much variation in quality can take place simply by varying M. This particular R came from table 7b. Both M's are admissible in the sense of §5 and [14, eq. (2.6)]. The *left*-hand column represents the "final" R and is about average in quality. Cf. tables 7b and 8.

〖The agreement between the c_n's is consistent with our basic rule-of-thumb.〗

* * * * * *

To further illustrate $N = 5$, it may be useful to take a look at some typical "output" for the case of *even* R. We do so in the following example. The *contrast* between odd & even speaks for itself. 〖Cf. §6 items (a)-(d).〗

Example 15. $R = 48.244$ ($N=5$ / even).

$R = 48.244655$

k	$\frac{1}{2}(c_k' + c_k'')$	rough diff.
2	-2.0084	1E-3
3	.4828	6E-5
4	-2.0096	2E-3
5	$-.5300$	1E-3
6	.2072	2E-5
7	-1.4332	2E-3
8	.7030	2E-3
9	.4987	9E-5
10	2.2295	3E-3
20	.994	1E-1
25	.016	1E-2
30	.093	9E-2

type (.40,.45 ‖ .42,.47)

M = 47

$R = 48.244535$

k	$\frac{1}{2}(c_k' + c_k'')$	rough diff.
2	-2.0389	5E-3
3	.5085	3E-6
4	-2.0615	7E-3
5	$-.5457$	5E-3
6	.2035	9E-6
7	-1.4718	9E-3
8	.735	1E-2
9	.5158	3E-4
10	2.264	1E-2
20	1.028	5E-2
25	.038	8E-2
30	.030	2E-1

type (.40,.45 ‖ .42,.47)

M = 48

$R = 48.244524$

k	$\frac{1}{2}(c_k' + c_k'')$	rough diff.
2	-2.037	2E-2
3	.492	2E-2
4	-2.037	2E-2
5	$-.537$	4E-2
6	.208	4E-3
7	-1.462	6E-2
8	.717	9E-2
9	.504	1E-2
10	2.258	5E-2
20	.941	3E-2
25	$-.034$	8E-2
30	.000	4E-2

type (.35,.40 ‖ .37,.42)

M = 57

$R = 48.247737$ (!)

k	$\frac{1}{2}(c_k' + c_k'')$	rough diff.
2	-2.075	2E-2
3	.590	6E-6
4	-2.197	4E-2
5	$-.602$	8E-3
6	.186	3E-4
7	-1.593	2E-2
8	.869	1E-2
9	.592	3E-2
10	2.290	7E-3
20	.969	1E-1
25	.051	4E-2
30	.047	2E-2

type (.35,.40 ‖ .37,.42)

M = 58

Type $(.40,.45 \| .42,.47)$ *also* included $M = 49$; nothing even remotely resembling 48.244 was picked up there. Similarly for type $(.35,.40 \| .37,.42)$ and $M = 56$.

Output of this kind certainly does *not* give one any reason to hope that $R = 48.244^+$ is an (even) eigenvalue. Things are simply too unstable/fuzzy.

Indeed: (if anything) one almost has the feeling that one is looking at some kind of "resonance" of finite width.

Note that the quality is definitely better for type $(.40,.45 \| .42,.47)$ than for $(.35,.40 \| .37,.42)$. This *agrees* with our earlier comment in §6 about pseudo-residuals.

Examples of this kind were a real *nuisance* in our production runs. They appeared much more frequently than we originally hoped -- and succeeded only in wasting a great deal of CPU time.

$$* \ * \ * \ * \ * \ *$$

For our last two examples, we return to R odd.

<u>Example 16</u>. $R = 50.663316$ (N=7 / odd) .

k	$\tfrac{1}{2}(c_k' + c_k'')$	diff.	$\tfrac{1}{2}(c_k' + c_k'')$	diff.	$\tfrac{1}{2}(c_k' + c_k'')$	diff.
	R = 50.663316		R = 50.663316		R = 50.663318	
2	2.441062	1E-5	2.441055	2E-6	2.441106	1E-7
3	-3.564807	1E-5	-3.564802	2E-6	-3.564764	4E-7
4	.577171	4E-6	.577168	8E-7	.577182	5E-7
5	-2.738239	1E-5	-2.738236	1E-6	-2.738188	2E-6
6	-.257097	1E-6	-.257100	1E-7	-.257058	4E-7
7	-2.772581	8E-6	-2.772574	3E-7	-2.772610	4E-6
8	1.286206	4E-6	1.286204	8E-8	1.286201	4E-6
9	.797624	2E-6	.797628	7E-7	.797585	1E-8
10	-1.351261	5E-6	-1.351259	1E-6	-1.351236	1E-6
15	.469160	1E-7	.469157	2E-6	.469193	6E-6
20	.415056	1E-6	.415059	2E-7	.415008	2E-6
25	-1.223638	9E-6	-1.223635	1E-5	-1.223658	5E-5
30	.16563	1E-5	.16563	2E-5	.1657	2E-4
40	.89907	4E-4	.89955	1E-4	.8978	2E-3
50	-2.0226	2E-3	-2.0237	5E-3	-2.034	3E-2
60	-3.25	0.24	-3.00	0.36	-2.85	0.40
type (.30,.35 $\|$ .32,.37) M = 75,76,77 resp.					3 tracks with large M - values	

The foregoing table provides yet another illustration of the "level-of-changes" that are appropriate for a *true* R when M is varied.

The most interesting feature of this example (*apart* from the change of group!!) is the relatively large size of M. Though some degradation in $|c_k' - c_k''|$ level *may* have occurred, it is scarcely noticeable. The overall state-of-affairs remains eminently satisfactory... [The figures shown are *typical* for this R-range.]

Incidentally: note that the middle column (and not the right) has been taken to be the "final R" in table 7d.

[We are not overly distressed by the apparent breakdown in our rule-of-thumb (concerning c_n-error) in the right column. That rule was only heuristic and was mainly intended for use in connection with (groups of) *semi*final R-values.

Given the size of M, it is entirely possible that R = 50.663318 is a bit "off" due to some kind of local static. Cf. [14] and paragraphs 4-5 following §6(d).]

Example 17.   R = 80.838014 (N=5 / odd) .

	R = 80.8380137			R = 80.8380141	
k	$(c_k' + c_k'')$	diff.	k	$(c_k' + c_k'')$	diff.
2	−4.62176	3E−5	2	−4.62174	8E−5
3	.95914	7E−6	3	.95915	2E−5
4	.17617	3E−6	4	.17615	2E−5
5	.93967	3E−7	5	.93969	2E−6
6	−.40911	3E−5	6	−.40910	3E−5
7	−1.95823	2E−5	7	−1.95825	8E−6
8	−1.62584	2E−6	8	−1.62583	3E−5
9	.85828	4E−5	9	.85822	4E−5
10	3.98313	6E−5	10	3.98306	1E−4
15	−1.83324	4E−5	15	−1.83321	6E−5
20	−.02789	9E−6	20	−.02789	1E−5
25	−2.25775	5E−5	25	−2.25771	4E−5
30	1.03086	4E−5	30	1.03083	3E−5
35	1.36517	9E−5	35	1.36511	3E−4
40	1.28698	2E−5	40	1.28694	1E−4
45	−6.04903	6E−5	45	−6.0489	2E−4
50	−1.4443	1E−4	50	−1.4433	3E−3
55	.304	1E−2	55	.58	0.57

type (.45,.50 ‖ .47,.52)		left M = 65	2 tracks with
		right M = 67	large R & M

The quality in this example is (still) quite good *considering* the size of R and M. The agreement between the c_n's is consistent with our basic rule-of-thumb, while the values of *diff* are just about average. Cf. table 8.

Incidentally, in table 7c, the entries $\{79.965505 , 80.497242 , 80.737485\}$ each appear for *only one* value of M. In the other 7 cases: the difference between the corresponding R-values never exceeds 8×10^{-7}.

§8. <u>Some</u> <u>Remarks</u> <u>about</u> <u>Pseudo</u> <u>Cusp</u> <u>Forms</u>. The frequent occurrence of unstable R-values of weak overall quality in our (even) production runs is a matter of some concern.

As mentioned earlier, these R-values appear to represent some kind of residual effect from the pseudo cusp forms. Cf. §§ 3, 6(near the end), and example 15.

In this section, we give a report on some experiments we did to try to clarify this matter. ([1])

Our discussion will be largely informal. [Its primary aim is simply to outline the main ideas. In particular: in connection with linear algebra, we shall generally be content to proceed with (only) rather heuristic estimates for any error terms.]

We continue to use the algorithm in §2.

Let $\rho \equiv \exp(\pi i/N)$. One of the most obvious ways of investigating unstable R-values is to examine their dependence on $\{z_j\} \cup \{w_j\}$. In §2, we basically assumed that z_j & w_j belonged to $\mathcal{F}_N$. We retain this assumption here for reasons of simplicity.

As shown in §3, the *unsafe* portion of $\mathcal{F}_N$ is the region

$$\mathcal{U}_N \equiv \mathcal{F}_N \cap \{ \operatorname{Im} E(z) > \operatorname{Im}(\rho) \}$$
$$\equiv \{ |x| < \tfrac{1}{2}\mathcal{L} \} \cap \{ \sqrt{1-x^2} < y < \frac{1 + \sqrt{1-(xS)^2}}{S} \}$$

where $\mathcal{L} = 2\cos(\frac{\pi}{N})$ and $S = 2\sin(\frac{\pi}{N})$. Keeping z_j & w_j inside $\mathcal{U}_N$ *maximizes* the risk of picking up pseudo cusp forms. ([2])

([1]) The group $\mathfrak{C}_N$ can be *either* arithmetic or nonarithmetic.

([2]) (Recall figure 1 in § 5.)

In line with these remarks, *our basic idea* was to simply run a variety of
unsafe experiments -- and to compare the *output* with that of our production
runs. Perhaps some pattern would be detected...

The batches we used were mainly of 2 types:

(A) those where the points z_j were taken along

$$y = \sqrt{1 - x^2} + \varepsilon_0$$

with x_j evenly distributed in $(0, \frac{1}{2}\mathcal{L})$; similarly for w_j ;

(B) those where the z_j were placed on various (other) configurations in $\mathcal{F}_N$
with only a certain proportion of the z_j actually lying inside $\mathcal{U}_N$; similarly
for w_j .

In both cases, we explicitly allow N to be 3 .

Batches of type (B) are called "hybrid" for the obvious reason.

In (A), the parameter ε_0 is kept very small. This *ensures* that $\{z_j\} \cup \{w_j\} \subseteq \mathcal{U}_N$
and that (2.4) approximates a Neumann boundary condition at z_j and w_j . ([3])

Taking ε_0 *too* small typically leads to excessive floating point errors in (2.5).
Some caution is definitely necessary. For modest M , we'll generally take ε_0 to be
either 10^{-3} or 10^{-4} .

To avoid difficulties at ρ , it is also useful to impose the further restriction
that z_j & w_j have real part $\leq \frac{\mathcal{L}}{2}(1 - \frac{1}{M})$.

- - - - - -

Prior to looking at any kind of experimental output, it is important to identify
several factors which play a key role in "controlling" what the machine actually sees.

We focus initially on case A.

In the following discussion: let Φ represent a generic (even) pseudo cusp form
having singularities in a *tiny* neighborhood $\mathcal{N}$ of ρ . Assume that Φ is unit-
normalized and that $\Delta\Phi + (\frac{1}{4} + R^2)\Phi = 0$.

At this stage, the number R is completely arbitrary. Cf. (3.3)(3.4).

The *extent* to which Φ is picked up by the machine hinges, at least in part, on
the "collective smallness" of its tail (after M terms) at the points z_j , w_j , $E(z_j)$,
$E(w_j)$. Cf. (2.4).

([3]) Initially at least all our jobs are even.

Let $\| \text{tail} \|$ be some type of *quantitative* measure for this smallness.

⟦Bear in mind here that each $I_n(z,R)$ is really a *difference*. For this reason, in defining $\| \text{tail} \|$, it would certainly seem natural to *re*scale things by dividing through by ε_o (just like a normal derivative). We tacitly assume that this has been done. Compare §2 footnote 1.⟧

The *other* important ingredient is the conditioning-level (or, more properly, conditioning behavior) of the matrices:

(8.1) $$\left[I_n(z_j', R) \right] \quad , \quad \left[I_n(w_j', R) \right] \qquad \bullet$$

Cf. (2.5) and [9,pp.25,72(II)].

It is necessary to introduce quantitative assessments $\{ C_z , C_w \}$ for these levels which *properly* reflect the enormous range of magnitudes typically found in (8.1).

To this end: we first observe that the matrices in (8.1) have *columns* which eventually scale (downwards) something like a geometric progression *wrt* n . Cf. [7]. In situations like this, it is generally expedient to make a judicious change-of-variable. Cf. [9,p.73(middle)]. Accordingly: we propose to set

(8.2) $$c_k \equiv s_k^{-1} \xi_k s_l \quad ,$$

where s_ℓ denotes some type of "proximate" order for $I_\ell(z_j, R)$ & $I_\ell(w_j, R)$. [4] This (in turn) leads to:

(8.3) $$\left[I_n(z_j', R)/s_n \right] \quad , \quad \left[I_n(w_j', R)/s_n \right] \qquad \bullet$$

The hope is that (8.3) will exhibit better conditioning than (8.1).

Let

(8.4) $$a' \xi = b' \quad , \quad a'' \xi = b''$$

be the counterpart of (2.5'),(2.5") . By construction: the entries in a', b', a'', b''

---

[4] The number s_ℓ is best regarded as a kind of norm.

are essentially $O(1)$ [i.e. *much* more level than before] . [5]

At the same time: observe that $a\dot{\xi} = \ell + t$, $a\xi = \ell$, and $\|t\| \overset{\leq}{=} \delta\|\ell\|$ automatically yield

$$(8.5) \qquad \frac{\|\dot{\xi} - \xi\|}{\|\xi\|} \overset{\leq}{=} \|a^{-1}\| \cdot \|a\|\delta \equiv K(a)\delta \quad .$$

The best cases have $K(a)$ modest and δ very small. [6] In practice, one naturally seeks to identify $\|t\|$ with $\|tail\|$.

In any event, we now take:

$$(8.6) \qquad C_z \equiv K(a') \quad , \qquad C_w \equiv K(a'') \quad .$$

From a *heuristic* standpoint, the overall situation is rather clear. In order for $\overline{\Phi}$ to be detected by the machine, it is basically necessary for something like

$$(8.7) \qquad C_z \|tail\| \ll 1 \quad , \qquad C_w \|tail\| \ll 1$$

to hold. [7]

Incidentally: note that the matrices in (8.1) & (8.3) are functions of N as well as M , R , z_j , w_j .

What is *not at all* clear apriori is (precisely) which R - values will show up -- and how sharply.

Part of the difficulty stems from the fact that $\|tail\|$ and C_z , C_w are "locked" in a kind of dynamic competition.

Indeed: consider the effect of pushing M [upwards] significantly beyond the initial range of values associated with (5.1).

On the *one* hand: this forces the points z_j to lie closer together -- which presumably causes C_z to deteriorate. Similarly for C_w .

On the *other* hand, raising M tends to *reduce* the tail size for each fixed

---

[5] A *certain amount* of variability (or "spread") will continue to exist in a' & a'' , particularly for large M .   Cf. [7].

[6] See [49,p.194(4.5)] for an alternate (more probabilistic) version of (8.5).

[7] We use the word *"basically"* because (8.7) is admittedly an *over*simplification. Among other things, we've failed to take into account the "fuzziness" which is intrinsic to both R & I_n . For present purposes, however, the main idea is adequately expressed by (8.7). [In our actual jobs, $R \overset{\leq}{=} 25$ and $M \overset{\leq}{=} 30$] .

$z \in \mathcal{U}_N$.   ⟦The effect on $\|\text{tail}\|$ is *not* so clear, however, because some points z_j will now lie even closer to $\mathcal{n}$. ⟧

The problem is clearly one of *several* variables -- and is a natural candidate for some empirical tests.

The following points ⟦re: (A)⟧ should also be kept in mind.

(i) In principle at least, the matrices (8.1) have nothing to do with automorphic forms. Their conditioning-level *might* be perfectly fine even for relatively large M. ([8]) In *that* case: one would be inclined to say that (for given R) there can't exist a wide variety of distinct Φ with abnormally small $\|\text{tail}\|$. Otherwise (2.5') would admit too many solutions.

This state-of-affairs would also seem to say that the bigger system (2.5')(2.5") should lead to a relatively sharp, discrete R - listing.

(ii) There is, of course, the *other* extreme. Namely: that $\mathcal{C}_z$ & $\mathcal{C}_w$ might tend to become *uniformly bad* (already for more moderate M). As emphasized above, these numbers *also* depend on N. Though any kind of rigorous prediction seems out-of-the-question, there is some *hint* that problems (of this kind) do indeed occur as N ↗ . This comes from (3.2). Note that:

$$exp\left[\pi \tan\left(\tfrac{\pi}{N}\right)\right] = \left\{ \begin{array}{ll} 230.765 & for \ N = 3 \\ 23.141 & for \ N = 4 \\ 9.801 & for \ N = 5 \\ 6.134 & for \ N = 6 \\ 4.540 & for \ N = 7 \end{array} \right\} \quad . \qquad ([9])$$

A generic Φ on $\mathfrak{C}_7$ will thus tend to have significantly smaller coefficients than one on $\mathfrak{C}_3$ ⟦at least for large n ⟧ . This suggests ([10]) that it becomes quite a bit easier to manufacture "relatively small" Φ (for given R) *once* N starts to increase. ([11])

To the extent that the *format* (3.3) is quite flexible, there appears to be some danger that $\mathcal{C}_z$, $\mathcal{C}_w$ will deteriorate.

(iii) In a similar way: artificially keeping z_j & w_j away from ρ opens the door to a *wider* class of competing Φ , hence to a greater likelihood that any particular R admits "too many" relatively small Φ . This means that the chances of encountering *bad* $\mathcal{C}_z$, $\mathcal{C}_w$ are that much greater as well.

⟦It seems difficult to make any kind of more precise prediction, however.⟧

([8]) (at least for *most* R)
([9]) And *recall* the power series analogy in [13,p.25(remark 4.11)] !
([10]) especially if the expansions (2.1) are treated as sums of random variables
([11]) Compare: [18,p.134(II),131(C)]. ⟦The phrase "relatively small Φ " is *short* for saying that $\|\text{tail}(\Phi)\|$ is relatively small.⟧

(iv) One prediction that *does* seem fairly safe is the following. Consider a given R-range with fixed M. Assume that C_z, C_w remain OK. Then: R-values associated with *pseudo* Φ should tend to be *less* accurate (in general) than those associated with true φ .

This simply reflects the fact that $\|\text{tail}(\Phi)\|$ will typically be much bigger than $\|\text{tail}(\varphi)\|$ (at least until both get down to the level of machine noise). Once *that* happens, or the conditioning goes, all bets are off...

— — — —

To some extent at least, analogous remarks hold for "hybrid" batches.

— — — —

We now turn to the *results*.

N = 3		
8.039731*	8.03973	8.03961
11.249201*	11.24920	11.24912
13.779752	13.77975	13.77975135*
14.134739*	14.13472	14.13460
15.704619*	15.70462	15.70461
17.738564	17.73856	17.7385634*
18.262036*	18.26197	18.26196
19.423482*	19.42348	19.42348
20.45571*	20.45575	20.45566
21.02204*	21.02204	21.02206
21.315796*	21.31580	21.31580
22.785909	22.78591	22.785909*
24.05944*	24.05940	24.05936
24.112353*	24.11235	24.11235
25.01076*	25.01084	25.01075
type A , $\varepsilon_0 = 10^{-4}$, M = 20,15,M_3 respectively		
M_3 = an *appropriate* smaller value ala (5.1)		
* denotes the case with best differences		
H1 = .025 ; H2 = .001 ; H3 = 10^{-6}		
$0 \leq R \leq 25.125$	EVEN R	Pseudo + True

TABLE   13

13.779751	14.134725	8.039737	6.020949
17.738563	21.022040	11.249206	10.243770
19.423481	25.010858	15.704619	12.988098
21.315796		18.261997	16.342607
22.785909		20.455771	18.291993
24.112353		24.059415	21.450611
			23.278377

column 1 : true R_n for $N = 3$		even R
column 2 : zeros of $\zeta(s)$		$s = \frac{1}{2} + iR$
column 3 : zeros of $L(s, \chi_{-3})$		$\chi_d(n) = \left(\dfrac{d}{n}\right)$
column 4 : zeros of $L(s, \chi_{-4})$		

TABLE   14

Some comments on tables 13 & 14. First of all, observe that Table 13 partly *duplicates* the earlier result of [11] and [16] . The pseudo R - values (appear to) correspond to (3.3) with $Q = 1$, $\tau_1 = \rho$. [12]

Two things are particularly striking:

(a) the accuracy;

(b) the fact that no other Φ appear.

Bear in mind here that (since $M \stackrel{<}{=} 20$) there certainly *does* exist a small neighborhood $\mathcal{N}$ of ρ which is admissible for current Φ - purposes.

On the basis of (a), one is inclined to speculate that the conditioning levels $\mathcal{C}_z , \mathcal{C}_w$ must (generally) remain fairly good, at least out to $M = 20$. Cf. remark (i).

On the basis of (b) : one is inclined to say that *the (unit-normalized) pseudo cusp forms associated with* $G_s(z;\rho)$ *must somehow be* distinguished [*at least locally*] *by the abnormal smallness of their* $\|tail\|$.

[In table 13, it is also interesting that there seems to be no residual effect from $G_s(z;i)$. There the relevant R would need to correspond to $\mathfrak{J}(s)L(s, \chi_{-4}) = 0$.]

— — — — — — —

In the absence of a careful numerical analysis of the linear algebra, it's a bit difficult to assign a precise level of reliability to the c_k associated with any particular Φ in table 13.

——————

[12] since $E(\rho ;s) \equiv 3\left(\frac{1}{2}\sqrt{3}\right)^s \dfrac{\mathfrak{J}(s) L(s, \chi_{-3})}{\mathfrak{J}(2s)}$

The *extremal property* suggested by (b) is "attractive enough" that one would *at least* want to push 14.134739 a bit further.

In retrospect: it would have been better if some form of our original CRAY code had been implemented in double-precision.

In order to proceed solely on the basis of the existing code, it is necessary to use reasoning which is partly heuristic.

The idea is as follows.

The algorithm in §2 was [fortunately] implemented in such a way that certain information *intermediate* to the derivation of $\{c_k', c_k'', R\}$ was always retained in the final listing.

This information included the local velocities v_k', v_k'' and velocity fluctuations d_k', d_k'' for the coefficients c_k', c_k'' [of the given Φ (or φ_n)] at the level of H2 .

Approximate values for the *scaling* factors s_k (over the entire R - range) were generally computed at the very start -- in a separate program.

In most cases, the R - values that were finally obtained were "stable" in the sense that:

$$(8.8) \qquad \left|\frac{d_k'}{v_k'}\right| \ll 1 \quad , \quad \left|\frac{d_k''}{v_k''}\right| \ll 1 \qquad for\ every\ k \lesssim M \ .$$

[Typical values for $R = 11.249201$, 14.134739 , 15.704619 range anywhere from *.002* to *.006* .]

Ratios of this size tell us that *linear* interpolation of c_k' & c_k'' is reasonable (near the given R - value) at least out to a "distance" of several H2 intervals.

This type of smoothness also provides further evidence (albeit *indirect*) that the numbers $\mathcal{C}_z$, $\mathcal{C}_w$ do indeed remain relatively modest over the given R -neighborhood. In particular: no zeros of $\det(a')$ or $\det(a'')$ can be present.

For ease of notation, we now let R_E denote the exact value of R (for Φ or φ_n) . Let R_M be the value proposed by the machine. In situations like (8.8), it is natural to proceed using the following heuristic representation:

$$(8.9) \qquad c_k'(R) = c_k + v_k'(R - R_E) + \frac{s_1}{s_k}\, m \|tail\| \,\theta_k'$$

$$c_k''(R) = c_k + v_k''(R - R_E) + \frac{s_1}{s_k}\, m \|tail\| \,\theta_k'' \qquad \bullet$$

Here m is a modest constant indicating the approximate size of $\|(a')^{-1}\|$, $\|(a'')^{-1}\|$ and θ_k' , θ_k'' are suitable numbers in [-1,1] . Note that θ_k' & θ_k'' are *independent* of R . (Indeed: let $R = R_E$.) $(^{13})$

The terms $\mathcal{M} \|tail\| \theta'_k$, $\mathcal{M} \|tail\| \theta''_k$ correspond to components in an appropriate $a^{-1}t$. As such: the sizes of θ'_k , θ''_k will partly reflect the "shape" of a^{-1} . Collectively, of course, one automatically has:

(8.10) $\dfrac{1}{K(a')} \stackrel{<}{\approx} \max_k |\theta'_k| \stackrel{<}{=} 1$, $\dfrac{1}{K(a'')} \stackrel{<}{\approx} \max_k |\theta''_k| \stackrel{<}{=} 1$

or something very similar (depending on which norm we use).

The problem here is obvious: there are *too many* unknowns.

Observe, however, that:

$$c'_k(R_M) - c''_k(R_M) = (v'_k - v''_k)(R_M - R_E) + \frac{s_1}{s_k}(\theta'_k - \theta''_k)\mathcal{M}\|tail\| \quad .$$

The batches $\{z_j\}$ and $\{w_j\}$ are intermingled. And, as we've just seen, the entries θ'_k & θ''_k arise from matrix multiplications ala $a^{-1}t$.

Statistically speaking, then, it is not too unreasonable to *expect* that

$$\alpha_k \equiv \frac{\theta'_k}{|\theta'_k - \theta''_k|} \quad , \quad \beta_k \equiv \frac{\theta''_k}{|\theta'_k - \theta''_k|}$$

will [typically] be fairly modest in size. [Keep in mind that $|z_j - w_j| \approx \dfrac{const}{M}$.]

Let:

(8.11) $\ell'_k = c'_k(R_M) + v'_k(R_E - R_M)$, $\ell''_k = c''_k(R_M) + v''_k(R_E - R_M)$ •

A quick computation shows that:

$$\ell'_k - \ell''_k = \frac{s_1}{s_k}(\theta'_k - \theta''_k)\mathcal{M}\|tail\|$$

(8.12) $c_k = \ell'_k - \alpha_k|\ell'_k - \ell''_k|$, $c_k = \ell''_k - \beta_k|\ell'_k - \ell''_k|$ •

$(^{13})$ Since the I_n are only computed to finite precision, the term $\|tail\|$ must be understood to include a certain amount of intrinsic fuzz. Cf. [9,p.25 near eq.(2.5.2)] and vector t near (8.5). This remark is particularly important if R_E corresponds to a true eigenvalue.

We like to think of ℓ_k', ℓ_k'' as defining a kind of *expected value* for c_k [in terms of R_E and the machine output]. In view of (8.12), this is reasonable so long as:

$$\frac{|\ell_k' - \ell_k''|}{|\ell_k'|} \ll 1 \qquad \bullet$$

To handle cases where R_E is *not* known, assume [next] that $|R_E - R_M| \lesssim h$. In this case, equation (8.12) yields:

$$c_k = c_k'(R_M) + O(1)(1+|\alpha_k|)\left[\,|c_k'(R_M)-c_k''(R_M)| + h(|v_k'|+|v_k''|)\,\right]$$

$$c_k = c_k''(R_M) + O(1)(1+|\beta_k|)\left[\,|c_k'(R_M)-c_k''(R_M)| + h(|v_k'|+|v_k''|)\,\right]$$

with implied constants which do not exceed **1** .

Our (statistical) hypothesis about α_k, β_k will *then* yield:

$$(8.13) \quad c_k = c_k'(R_M) + \tilde{O}(1)\left[\,|c_k'(R_M)-c_k''(R_M)| + h(|v_k'|+|v_k''|)\,\right]$$

$$c_k = c_k''(R_M) + \tilde{O}(1)\left[\,|c_k'(R_M)-c_k''(R_M)| + h(|v_k'|+|v_k''|)\,\right]$$

where $\tilde{O}(1)$ means a term which is *usually* bounded. In floating point notation, this becomes:

$$(8.14) \qquad c_k = c_k'(R_M)\left[\,1 + \tilde{O}(1)\eta_k'\,\right] \quad , \quad c_k = c_k''(R_M)\left[\,1 + \tilde{O}(1)\eta_k''\,\right]$$

where

$$\eta_k' \equiv \frac{|c_k'(R_M)-c_k''(R_M)| + (|v_k'|+|v_k''|)h}{|c_k'(R_M)|} \quad , \quad \eta_k'' \equiv \frac{same\ expression}{|c_k''(R_M)|} \qquad \bullet$$

Equations (8.13) & (8.14) presuppose the *stability* condition (8.8). In cases where these (curvature) ratios start to become a bit larger, it is certainly possible to include *higher-order* terms in (8.9).

It is also interesting to compare (8.13) with the old rule-of-thumb in §7. Equation (8.13) is definitely the more conservative of the two. ([14])

To round things out, we need to explain how to obtain (at least) a ballpark estimate for $\|\text{tail}\|$. This is very easy. In fact:

$$\frac{s_k}{s_I}\frac{|\ell_k'-\ell_k''|}{|\theta_k'-\theta_k''|} \;=\; \mathfrak{M}\|\text{tail}\| \;=\; \left|\frac{\alpha_k}{\theta_k'}\right|\frac{s_k}{s_I}|\ell_k'-\ell_k''|$$

$$\frac{1}{2}\frac{s_k}{s_I}|\ell_k'-\ell_k''| \;\le\; \mathfrak{M}\|\text{tail}\| \;\le\; \widetilde{O}(1)\frac{1}{|\epsilon_k'|}\frac{s_k}{s_I}|\ell_k'-\ell_k''|$$

$$\frac{1}{2}\max_k\left[\frac{s_k}{s_I}|\ell_k'-\ell_k''|\right] \;\le\; \mathfrak{M}\|\text{tail}\| \;\le\; \widetilde{O}(1)\min_k\frac{1}{|\theta_k'|}\cdot\max_k\left[\frac{s_k}{s_I}|\ell_k'-\ell_k''|\right]$$

$$\frac{1}{2}\max_k\left[\frac{s_k}{s_I}|\ell_k'-\ell_k''|\right] \;\le\; \mathfrak{M}\|\text{tail}\| \;\le\; \widetilde{O}(1)K(a')\cdot\max_k\left[\frac{s_k}{s_I}|\ell_k'-\ell_k''|\right]$$

$$\frac{1}{2\mathfrak{M}}\max_k\left[\frac{s_k}{s_I}|\ell_k'-\ell_k''|\right] \;\le\; \|\text{tail}\| \;\le\; \widetilde{O}(1)\|a'\|\cdot\max_k\left[\frac{s_k}{s_I}|\ell_k'-\ell_k''|\right]$$

by (8.10). The desired estimate is therefore:

(8.15) $$\|\text{tail}\| \;\approx\; \max_k\left[\frac{s_k}{s_I}|\ell_k'-\ell_k''|\right]$$.   ([15])

We stress that this equation presupposes good conditioning. [See also footnote 13.]

These heuristics now lead to the following table.

([14]) Though (8.13) has been derived for type A batches, there is little difficulty extending it to more general $\{z_j\}\cup\{w_j\}$. In this connection, note that the "tail" terms in (8.9) are best left in the form $\mathfrak{F}_k'$, $\mathfrak{F}_k''$.

([15]) For arbitrary batches, one obtains upper bounds for $\|\text{tail}\|'$, $\|\text{tail}\|''$ and a lower bound for $\|\text{tail}\|' + \|\text{tail}\|''$. The latter uses $\sigma_k \equiv \min\left(\frac{s_k'}{s_I'},\frac{s_k''}{s_I''}\right)$.

$$R = 14.134739$$

k	c_k	$|c_k/c_{k-1}|$	$(|v_k'| + |v_k''|)|\Delta R|$	$|c_k' - c_k''|$
2	4.5300	4.53	3E-4	4E-6
3	7.967	1.76	1E-3	6E-7
4	-1.717E+2	21.55	5E-2	4E-3
5	1.211E+4	70.53	7E+0	9E-1
6	-1.291E+6	1.066E+2	1E+3	4E+2
7	1.719E+8	1.332E+2	3E+5	1E+5
8	-2.62E+10	1.52E+2	6E+7	5E+7
9	4.37E+12	1.67E+2	2E+10	1E+10
10	-7.75E+14	1.77E+2	3E+12	6E+12
11	1.44E+17	1.86E+2	1E+15	1E+15
12	-2.8E+19	1.94E+2	2E+17	7E+17
13	5.5E+21	1.96E+2	1E+20	1E+20
14	-1.1E+24	2.0E+2	1E+22	9E+22
15	2.3E+26	2.1E+2	1E+25	1E+25

Tentative Coefficients for Φ on $\mathbb{C}_3$	
hopefully off only in the last digit!!	c_k based on:
Limiting ratio: 230.765	(8.13)

TABLE   15

Application of (8.15) yielded $\|\text{tail}\| \approx 10^{-2}$. Similar values were obtained for $R = 8.039731$, 11.249201, and 15.704619 . For $R = 13.779752$, on the other hand, we got 10^{-7}. Since the *real* $\|\text{tail}\|$ must be something like 10^{-36} , the number 10^{-7} serves mainly to indicate the *level* of intrinsic fuzz. ([16])

— — — — — — —

We also experimented ([17]) with restricting $\text{Re}(z_j)$ to lie in $(0, \frac{2}{5}\mathcal{L})$. This caused *both* the true & pseudo R – values to become distorted (and of lower quality) especially for larger M . Though this agrees with remark (iii), the exact cause for the "conditioning erosion" requires further investigation.

When $(0, \frac{2}{5}\mathcal{L})$ was replaced by $(\frac{1}{10}\mathcal{L}, \frac{1}{2}\mathcal{L})$, a similar deterioration occurred but mainly for pseudo R . This makes sense since $\|\text{tail}(\Phi)\|$ *should* tend to be a bit larger here.

— — — — — —

([16]) For $R = 13.779752$, it is tempting to seek a maximum likelihood estimate for R_E by equating ℓ_k' and ℓ_k'' . When this is done, one finds that $R_E = 13.77975135$ for *every* $k \lessapprox 20$. Needless to say, this represents a strong affirmation of the overall accuracy.

([17]) (*in* type A)

The situation for N = 4 and N = 6 was essentially similar to N = 3 except for the fact that the conditioning decay (wrt M) began quite a bit earlier than we would have liked, particularly for N = 6 . ⟦ Cf. remark (ii).⟧

The analysis in §4 (involving f_o) is easily modified to allow one to compute the Eisenstein series for $\mathfrak{C}_4$ and $\mathfrak{C}_6$ in terms of the one for PSL(2,**Z**) . In particular: one finds that

$$
(8.16) \qquad E_N(\rho;s) \;=\; \begin{cases} \dfrac{2}{1+\rho^s}\, E_3(i;s) & \text{for } N = 4 \\[2mm] \dfrac{2}{1+\rho^s}\, E_3[\tfrac{1}{2}+\tfrac{1}{2}i\sqrt{3}\,;s] & \text{for } N = 6 \end{cases} \Bigg\} \qquad .
$$

N = 4	
6.020951*	6.02094
8.922877*	8.92288
10.243783*	10.24378
10.920392*	10.92039
12.98809*	12.98818
13.779745*	13.77975
14.13477*	14.13478
14.685016*	14.68502

type A, $\mathcal{E}_{\bullet} = 10^{-4}$, M = 25, M_2 , resp.
M_2 = an *appropriate* smaller value ala (5.1)
* denotes the case with best differences

H1 = .025 ; H2 = .001 ; H3 = 10^{-6}		
$0 \overset{\le}{=} R \overset{\le}{=} 15.125$	EVEN R	Pseudo + True

TABLE   16

Table 16 shows excellent agreement with tables 9 and 14(columns 2,4). The only pseudo R - values that were detected appear to come from $G_s(z;\rho)$. It is *therefore* likely that the forms associated with $G_s(z;\rho)$ are ⟦ once again ⟧ distinguished by their abnormally small ‖tail‖ .

$R = 14.13477$

k	c_k	$\lvert c_k / c_{k-1} \rvert$	$(\lvert v_k' \rvert + \lvert v_k'' \rvert)\lvert \Delta R \rvert$	$\lvert c_k' - c_k'' \rvert$
2	2.7058	2.71	2E-4	1E-11
3	-2.0950	0.77	5E-4	4E-11
4	1.403	0.67	1E-3	2E-11
5	4.668	3.33	3E-3	4E-9
6	-1.772E+1	3.80	3E-2	8E-8
7	1.515E+2	8.55	6E-1	6E-6
8	-1.68E+3	1.11E+1	1E+1	3E-4
9	2.22E+4	1.32E+1	3E+2	2E-2
10	-3.30E+5	1.49E+1	6E+3	7E-1
11	5.3E+6	1.61E+1	2E+5	5E+1
12	-9.2E+7	1.7E+1	4E+6	5E+1
13	1.7E+9	1.8E+1	9E+7	2E+5
14	-3.2E+10	1.9E+1	2E+9	1E+7
15	6.3E+11	2.0E+1	6E+10	2E+9

Tentative Coefficients for $\overline{\Phi}$ on $\mathfrak{C}_4$

hopefully off only in the last digit	c_k based on:
Limiting ratio: 23.141	(8.13)

TABLE 17

* * * * * * * *

N = 6	
5.098742	17.73864
[8.039]	18.26177
9.743749	18.64744
11.249180	18.96264
11.346418	[19.22259]
11.889978	19.42348
13.135144	19.89611
13.779721	[20.38504]
[14.10768]	20.45588
14.134726	20.66476
14.626241	[20.94812]
15.704601	21.02168
15.799508	[21.07796]
[16.23239]	21.31581
16.27096	21.43464
16.73620	21.80713
17.50057	

type A , $\mathcal{E}_{\theta} = 10^{-3}$, M = 20 *only*		
the conditioning started to deteriorate beyond M = 20 [20 is admissible ala (5.1)]		
very weak R are contained in brackets		
H1 = .025 ; H2 = .001 ; H3 = 10^{-6}		
$0 \leqq R \leqq 22$	EVEN R	Pseudo + True

TABLE 18

Apart from the bracketed values, table 18 shows good-to-excellent agreement with tables 10 and 14(columns 2,3).

The entry [8.039] is more-or-less expected: 8.038861 and 8.039737 are rather close together. The *other* bracketed entries *may* represent pseudo R which are a little less minimal than $G_s(z;\rho)$. [This would be consistent with remark (ii). Such R might *even* have occurred at N = 3 had we run everything in double precision!!]

In this connection, it should also be noted that the last 6 bracketed entries stay surprisingly "intact" at M = 25 ; *i.e.* 14.10710, 16.23211, 19.22252, 20.38484, 21.07795 . [The *non*bracketed ones either tend to disappear, or be violently shifted.]

$$R = 14.134726$$

k	c_k	$|c_k/c_{k-1}|$	$(|v_k'| + |v_k''|)|\Delta R|$	$|c_k' - c_k''|$
2	1.808582	1.81	8E-7	1E-11
3	.06240	.03	7E-6	1E-11
4	-1.13363	18.17	8E-6	2E-11
5	-.67494	.60	8E-6	3E-11
6	1.85381	2.75	4E-6	4E-12
7	2.36786	1.28	7E-6	4E-9
8	-2.36325	1.00	6E-5	2E-8
9	5.0601	2.14	4E-4	1E-6
10	-11.120	2.20	2E-3	3E-5
11	32.10	2.89	1E-2	1E-3
12	-1.091E+2	3.40	9E-2	3E-2
13	4.266E+2	3.91	6E-1	7E-1
14	-1.76E+3	4.13	4E+0	2E+1
15	7.7E+3	4.38	3E+1	5E+2

Tentative Coefficients for Φ on $\mathfrak{C}_6$

hopefully off only in the last digit	c_k based on:
Limiting ratio: 6.134	(8.13)

TABLE 19

* * * * * * * * *

The results for N = 5 have a slightly different emphasis since the *zeros* of $E_5(\rho\,;s)$ are *not* known.

	N = 5	
		5.390832*
		8.222146*
		8.837643*
10.126609*	10.126614	10.126614
12.609817	12.610666*	12.610709
13.638862	13.638825*	13.638826
14.839693	14.841641*	14.841611
[14.91170]*	[14.91212]	[14.92521]
type A , $\mathcal{E}_o = 10^{-4}$, M = 25, 22, M_3 , respectively		
M_3 = an *appropriate* smaller value ala (5.1)		
the conditioning began to deteriorate beyond M = 25		
very weak R are contained in brackets		
* denotes the case with best differences		
H1 = .025 ; H2 = .001 ; H3 = 10^{-6}		
$0 \leqq R \leqq 15.125$	EVEN R	Pseudo

TABLE   20

Though the R - values are fairly sharp, there is a noticeable drop in quality (across the columns). This may simply mean that the conditioning is *already* starting to "go" around $19 \sim 22$. $(^{18})$

The following examples were prepared assuming that $|\Delta R| \cong 5 \times 10^{-5}$ holds in column 2. The fact that one obtains only 10 or 11 *usable* coefficients (out of 22) tends to support our remark about $19 \sim 22$.

$(^{18})$ A systematic comparison was made of the c_k - velocities in our type A and type $(\propto \|\, \beta\,)$ batches. Though the velocities were comparable for small k , those of type *A* were significantly bigger once k reached 10 or so. This would seem to indicate a fundamental *difference* in the overall conditioning-level 〖and suggests the need for a certain amount of caution〗.

R = 12.610666

k	c_k	$\lvert c_k/c_{k-1}\rvert$	$(\lvert v'_k\rvert + \lvert v''_k\rvert)\lvert\Delta R\rvert$	$\lvert c'_k - c''_k\rvert$
2	2.4933	2.49	2E-4	3E-11
3	-1.385	0.56	8E-4	7E-11
4	1.514	1.09	7E-4	3E-11
5	2.465	1.63	2E-3	2E-9
6	-2.57	1.04	9E-3	2E-8
7	8.38	3.26	8E-2	5E-7
8	-2.81E+1	3.35	7E-1	9E-6
9	1.40E+2	4.98	7E+0	8E-4
10	-8.0E+2	5.7	8E+1	7E-3
11	5.1E+3	6.4	9E+2	8E-1
12	-3.5E+4	6.9	1E+4	1E+1

Tentative Coefficients for Φ on $\mathfrak{C}_5$

hopefully off only in the last digit	c_k based on:
theoretical ratio: 9.801	(8.13)

TABLE 21

R = 14.841641

k	c_k	$\lvert c_k/c_{k-1}\rvert$	$(\lvert v'_k\rvert + \lvert v''_k\rvert)\lvert\Delta R\rvert$	$\lvert c'_k - c''_k\rvert$
2	1.3872	1.39	2E-4	4E-11
3	.2233	.16	5E-4	6E-11
4	-.3921	1.76	5E-4	2E-10
5	-.5303	1.35	2E-4	2E-10
6	2.287	4.31	1E-3	4E-9
7	.678	.30	5E-3	2E-8
8	-.62	.91	3E-2	4E-8
9	2.9	4.7	2E-1	1E-5
10	-1.5E+1	5.2	2E+0	1E-4
11	8.5E+1	5.7	2E+1	1E-2
12	-5.5E+2	6.5	2E+2	2E-1

Tentative Coefficients for Φ on $\mathfrak{C}_5$

hopefully off only in the last digit	c_k based on:
theoretical ratio: 9.801	(8.13)

TABLE 22

The differences in tables 21 & 22 are *unusually* strong.

In fact: table 22 has the interesting "property" that it can be easily mistaken for output associated with a *true* R (especially if columns 3 & 4 are erased). This example *underscores* the need for testing multiple batches as in §6.

There is little doubt that the *un*bracketed entries in table 20 come from

abnormally small $\overline{\Phi}$. To the extent that R is real, we are inclined to *conjec-*
ture that $\overline{\Phi}$ is simply (constant)$\cdot G_s(z;\rho)$.

Compare: [13,p.251(d)].

Things may change for larger R , since $E_5(\rho;s)$ presumably has many zeros
off the critical line.

* * * * * * * * *

Tables 16,18,20 can now be compared with various "hybrid" batches -- and with
our production runs.

The results for N = 5 were rather striking. Here is a sample:

"Hybrid" Batches

10.126602	10.126602	10.126602	10.126606	10.126603	10.126602
12.608098	12.609420	12.609415	12.609754	12.609538	12.609514
13.638659	13.638744	13.638746	13.638763	13.638755	13.638748
14.840951	14.840946	14.840971	14.841098	14.840957	14.841025
60% in $\mathcal{U}_5$	60% in $\mathcal{U}_5$	50% in $\mathcal{U}_5$	50% in $\mathcal{U}_5$	30% in $\mathcal{U}_5$	30% in $\mathcal{U}_5$
$9.875 \lesssim R \lesssim 15.125$		R - values with *best* differences			

TABLE   23a

Production Runs

	R	detected on	R	detected on
10.126609	(10.126)	0 of 9 tracks	(10.126)	0 of 9 tracks
12.610666	12.609901	8 of 9 tracks	(12.610)	0 of 9 tracks
13.638825	13.638891	7 of 9 tracks	13.640241	1 of 9 tracks
14.841641	14.842583	9 of 9 tracks	14.850131	3 of 9 tracks
from table 20	type (.50,.55 ‖ .52,.57)		type (.40,.45 ‖ .42,.47)	
	$9.875 \lesssim R \lesssim 15.125$		R - values with *least BAD* differences	

TABLE   23b

Sample Coefficients for   14.841[+]

c_2	1.3872	1.3886	1.3886	1.3886	1.3883	1.3886	1.3885	1.3852	1.3694
c_3	.2233	.2266	.2266	.2265	.2259	.2266	.2262	.2186	.1844
c_4	-.3921	-.3888	-.3888	-.3889	-.3895	-.3888	-.3892	-.3959	-.4360
c_5	-.5303	-.5289	-.5289	-.5290	-.5292	-.5289	-.5291	-.5272	-.5529
c_6	2.287	2.2952	2.2953	2.2950	2.2935	2.2951	2.2943	2.2947	2.0829
c_7	.678	.6495	.6495	.6501	.6553	.6501	.6526	.6648	-1.8784
c_8	-.62	-.4533	-.4530	-.4547	-.4842	-.4575	-.4702	-.5730	.8609
c_9	2.9	1.8053	1.8043	1.8046	1.9916	1.8381	1.9112	2.7123	-2.4414
c_{10}	-15	-6.7110	-6.7100	-6.6512	-7.8446	-6.9566	-7.3799	-13.7913	159.5259
from table 22	hybrid							$(.50 \sim .57)$	$(.40 \sim .47)$

TABLE   24

Tables 23 & 24 clearly demonstrate the residual effect from table 20.

In particular: note how the effect *weakens* as α and β recede from $\text{Im}(\rho)$.

It should also be recalled here that batches of type $(\alpha \| \beta)$ were originally introduced to obtain better conditioning in (2.5). Cf. the first few paragraphs of §5. This improvement is nicely illustrated by studying c_7, c_8, c_9, c_{10} in the penultimate column of table 24. [Bear in mind here that $|\sin(\frac{\pi}{5}) - .57|$ is *only* about .018 .]

Residual effects were *also* observed for N = 4 and 6 . Here, however, we didn't bother to run so many test cases.

For N = 4 , there was virtually no sign of Φ in the *production* runs. This is *not* too surprising given that:

$$\left| \sin(\tfrac{\pi}{4}) - .57 \right| \; = \; .137$$

and that the c_n are quite a bit larger here [than for N = 5] . Cf. §6 (para 2) and table 17.

We initially hoped that the production runs for N = 6 would reach the "standard" set by N = 5 (with regard to Φ). Unfortunately this did not happen. We suspect that this simply reflects a general worsening of the conditioning as N ↗ [much like remark (ii)] . In any event: numbers like 15.70270 and 18.26268 *did* appear [in type (.40,.45 \| .42,.47)] .

– – – – – – –

Before we can develop any kind of heuristic explanation for the observed residual effects, we need to examine tables 15, 17, 19, 21, 22 more closely. It is necessary to *decide* whether the relatively slow convergence of $|c_k/c_{k-1}|$ to $\exp[\pi\tan(\frac{\pi}{N})]$ is *intrinsic* -- or whether it's a manifestation of some sort of extremal behavior.

To this end: take $N = 3$ and assume that $G_s(z;\rho)$ is a pseudo cusp form with $s = \frac{1}{2} + iR$ and $R > 1$.

By virtue of [13,p.42], we know that

$$(8.17) \qquad G_s(z;\rho) = E(\rho;s)\frac{y^{1-s}}{1-2s} - \sum_{n\neq 0} F_n(\rho;s)y^{\frac{1}{2}} K_{s-\frac{1}{2}}(2\pi|n|y)e^{-2\pi inx}$$

for $\text{Im}(z) \geq \frac{1}{2}\sqrt{3}$. In particular:

$$(8.18) \qquad F_m(\rho;s)Y^{\frac{1}{2}} K_{iR}(2\pi|m|Y) = -\int_{-1/2}^{1/2} G_s[x+iY;\rho]e^{2\pi imx}dx$$

for any $Y \in [\frac{1}{2}\sqrt{3}, 1]$.

Since $G_s(z;\rho)$ is even, there is *no* loss of generality if we take m to be positive.

To get hold of $F_m(\rho;s)$, we need to estimate $G_s[x+iY;\rho]$. This is easily done using the methods of [13,pp.243-253, 317-319].

In fact, with $2 \leq a \leq 3$ and $z = x + iY$, we have:

$$G_s(z;\rho) = G_a(z;\rho) + [a(1-a) - s(1-s)]\int_{\mathcal{F}} G_a(z;\xi)G_s(\xi;\rho)\,d\mu(\xi)$$

$$\{\text{cf. } [13,p.252(\text{middle})]\}$$

$$G_s(z;\rho) = G_a(z;\rho) + O(R^2)\left\{\int_{\mathcal{F}} |G_s(\xi;\rho)|^2 d\mu(\xi)\right\}^{1/2} \qquad .$$

But:

$$\int_{\mathcal{F}} |G_a(w;\rho)|^2 d\mu(w) = \sum_n \frac{|\varphi_n(\rho)|^2}{[a(1-a)-\lambda_n]^2} + \frac{1}{2\pi}\int_0^\infty \frac{|E(\rho;\frac{1}{2}+it)|^2}{[a(1-a)-\frac{1}{4}-t^2]^2}dt$$

$$\int_{\mathcal{F}} |G_s(\xi;\rho)|^2 d\mu(\xi) = \sum_n \frac{|\varphi_n(\rho)|^2}{[s(1-s)-\lambda_n]^2} + \frac{1}{2\pi}\int_0^\infty \frac{|E(\rho;\frac{1}{2}+it)|^2}{[s(1-s)-\frac{1}{4}-t^2]^2}dt \qquad .$$

[This last equation makes essential use of the fact that $E(\rho;\frac{1}{2}+iR) = 0$.]

Let $0 < \delta < 1$ be chosen so that $(R-\delta, R+\delta)$ contains no R_n. We immediately deduce that:

$$\int_{\mathcal{F}} |G_s(\mathfrak{z};\rho)|^2 d\mu(\mathfrak{z}) = O\left(\frac{R^2}{s^2}\right) + O[f(R)]$$

where

$$(8.19) \qquad f(R) \equiv \frac{1}{R^2} \int_{|t-R|<1} \left| \frac{E(\rho; \frac{1}{2}+it)}{t-R} \right|^2 dt \qquad \bullet$$

For *arithmetic* $\mathfrak{C}_N$, integrals like (8.19) are easily estimated using techniques from analytic number theory.

In the present case, one simply writes

$$E(\omega; S) \equiv \frac{E_0(\omega; S)}{\mathfrak{J}(2S)}$$

and then applies (standard) Phragmén-Lindelöf techniques to bound $\frac{d}{dS}E_0(\omega;S)$ along $\mathrm{Re}(S) = \frac{1}{2}$. The contribution from $\mathfrak{J}(1+2it)$ is handled using [50,pp.42,53]. In this way:

$$f(R) = O(1) \qquad \bullet$$

Returning to (8.18), we see that:

$$(8.20) \qquad F_m(\rho;s) Y^{1/2} K_{iR}(2\pi|m|Y) = O\left(\frac{R^3}{s}\right)$$

uniformly wrt m .

It is now tempting to substitute various values of m & Y into (8.20) to gain [at least] *some* hold on $F_m(\rho;s)$. In doing so, bear in mind that the asymptotics of $K_{iR}(2\pi|m|Y)$ are controlled by [7].

Since $E(\rho; \frac{1}{2}+iR) = 0$, we automatically have $R > 8.039737$. In particular: $R > 2\pi$. This makes it clear that

$$(8.21) \qquad F_1(\rho;s) = \gamma \exp\left(\frac{\pi}{2}R\right) \quad,$$

where $|\gamma|$ is bounded from above by $\frac{1}{s}$ times a modest function of R .

At the same time: taking $Y \equiv \frac{1}{2}\sqrt{3}$ shows that

$$(8.22) \qquad F_m(\rho;s) = O\left[\frac{R^3}{s}\right] |m|^{1/2} \exp\left(\pi|m|/\sqrt{3}\right)$$

for $m \gg \frac{R}{\pi\sqrt{3}}$.

Estimates similar to (8.21) and (8.22) can obviously be derived for pseudo cusp forms like (3.3) and for $N = 4, 6$.

Nonarithmetic groups can also be included here if the analog of (8.19) has a good bound.

For $N = 5$ and modest R, we certainly don't expect any problems. ([19])

At this point, it is natural to ask whether a better estimate for $F_m(\rho;s)$ can possibly be derived by using the fact that we know the exact singularity of $G_s(z;\rho)$ at $z = \rho$. In some sense: we want to apply a Tauberian theorem to (8.17).

Without undertaking any kind of rigorous development, we simply remark that a little manipulation [with the precise asymptotics of $K_{iR}(X)$] *suggests* that $F_m(\rho;s)$ should look something like:

$$(8.23) \qquad (constant) \, (-1)^m \, \frac{e^{R g (X_m/R)}}{\sqrt{m}}$$

for large m, where

$$g(t) = \sqrt{t^2 - 1} + \arcsin\left(\frac{1}{t}\right) \qquad and \qquad X_m = 2\pi m \frac{Im(\rho)}{\chi} = \pi m \tan\left(\frac{\pi}{N}\right) \quad .$$

An alternate formulation would be:

$$(8.23') \qquad (constant) \, (-1)^m \, \frac{e^{R g (X_m/R)}}{\sqrt[4]{X_m^2 - R^2}} \quad .$$

In checking these claims, it is helpful to keep in mind that:

$$\left\{ \begin{array}{l} g(t) = t + \frac{1}{2t} + O(t^{-3}) \quad for\ large\ t \\[2mm] g'(t) = \frac{1}{t}\sqrt{t^2 - 1} \\[2mm] g(t)/t \quad is \quad monotonic \quad decreasing \quad for\ t \geq 1 \end{array} \right\} \quad .$$

For the unit normalized Φ which corresponds to $G_s(z;\rho)$, we should therefore have:

$$(8.24) \qquad c_m \cong \frac{constant}{\gamma} (-1)^m \frac{e^{R[g(X_m/R) - \frac{\pi}{2}]}}{\sqrt{m}}$$

([19]) Compare [48, p.627(3)].

or

$$(8.24') \qquad c_m \cong \frac{constant}{\gamma} (-1)^m \frac{e^{R[g(x_m/R) - \frac{\pi}{2}]}}{\sqrt[4]{x_m^2 - R^2}} \qquad \bullet$$

Cf. (8.21).

If one is lucky, these formulas will begin to hold for m just slightly bigger than the "threshold" value of $\frac{R}{\pi tan(\frac{\pi}{N})}$.

Let:

$$(8.25) \qquad \sigma_m = \frac{(-1)^m}{\sqrt{m}} e^{R[g(x_m/R) - \frac{\pi}{2}]}$$

$$(8.25') \qquad \tau_m = \frac{(-1)^m}{\sqrt[4]{x_m^2 - R^2}} e^{R[g(x_m/R) - \frac{\pi}{2}]} \qquad \bullet$$

We can now try to fit (8.24) + (8.24') to tables 15 , 17 , 19 , 21 , 22 and see what happens.

The *hope* is that σ_m/c_m & τ_m/c_m will quickly stabilize at some modest constant (as m ↗).

Here are the results.

$$\boxed{R_M = 14.134739 , \quad N = 3}$$

k	$\left\lvert c_k/c_{k-1} \right\rvert$	$\left\lvert \sigma_k/\sigma_{k-1} \right\rvert$	$\left\lvert \sigma_k/c_k \right\rvert$	$\left\lvert \tau_k/\tau_{k-1} \right\rvert$	$\left\lvert \tau_k/c_k \right\rvert$
4	21.55	31.00	.22	25.15	.11
5	70.53	74.84	.24	70.60	.11
6	1.066E+2	1.098E+2	.24	1.069E+2	.11
7	1.332E+2	1.354E+2	.25	1.334E+2	.11
8	1.52E+2	1.54E+2	.25	1.52E+2	.11
9	1.67E+2	1.68E+2	.25	1.67E+2	.11
10	1.77E+2	1.78E+2	.25	1.77E+2	.11
11	1.86E+2	1.86E+2	.25	1.85E+2	.11
12	1.94E+2	1.92E+2	.25	1.91E+2	.11
13	1.96E+2	1.97E+2	.25	1.96E+2	.11
14	2.0E+2	2.0E+2	.25	2.0E+2	.11
15	2.1E+2	2.0E+2	.24	2.0E+2	.11

threshold $\cong$ 3   TABLE 25a

$$R_M = 14.13477 \quad , \quad N = 4$$

k	$|c_k/c_{k-1}|$	$|\sigma_k/\sigma_{k-1}|$	$|\sigma_k/c_k|$	$|\tau_k/\tau_{k-1}|$	$|\tau_k/c_k|$
6	3.80	5.42	.22	4.40	.15
7	8.55	8.86	.23	8.24	.15
8	1.11E+1	1.15E+1	.24	1.11E+1	.15
9	1.32E+1	1.35E+1	.24	1.32E+1	.15
10	1.49E+1	1.51E+1	.24	1.48E+1	.15
11	1.61E+1	1.63E+1	.25	1.61E+1	.15
12	1.7E+1	1.7E+1	.25	1.7E+1	.14
13	1.8E+1	1.8E+1	.24	1.8E+1	.14
14	1.9E+1	1.9E+1	.24	1.9E+1	.14
15	2.0E+1	1.9E+1	.23	1.9E+1	.13

threshold $\cong$ 5   TABLE 25b

$$R_M = 14.134726 \quad , \quad N = 6$$

k	$|c_k/c_{k-1}|$	$|\sigma_k/\sigma_{k-1}|$	$|\sigma_k/c_k|$	$|\tau_k/\tau_{k-1}|$	$|\tau_k/c_k|$
9	2.14	1.91	.14	1.28	.15
10	2.20	2.66	.17	2.38	.16
11	2.89	3.21	.19	3.02	.17
12	3.40	3.63	.20	3.50	.17
13	3.91	3.96	.21	3.86	.17
14	4.13	4.24	.21	4.16	.17
15	4.38	4.46	.22	4.40	.17

threshold $\cong$ 8   TABLE 25c

$$R_M = 12.610666 \quad , \quad N = 5$$

k	$|c_k/c_{k-1}|$	$|\sigma_k/\sigma_{k-1}|$	$|\sigma_k/c_k|$	$|\tau_k/\tau_{k-1}|$	$|\tau_k/c_k|$
8	3.35	4.35	.26	4.01	.20
9	4.98	5.33	.27	5.10	.20
10	5.7	6.1	.29	5.9	.21
11	6.4	6.6	.30	6.5	.21
12	6.9	7.1	.31	7.0	.22

threshold $\cong$ 6   TABLE 25d

$$R_M = 14.841641 \quad , \quad N = 5$$

k	$|c_k/c_{k-1}|$	$|\sigma_k/\sigma_{k-1}|$	$|\sigma_k/c_k|$	$|\tau_k/\tau_{k-1}|$	$|\tau_k/c_k|$
8	.91	2.87	2.33	2.29	2.02
9	4.7	4.1	2.03	3.7	1.62
10	5.2	5.0	1.96	4.8	1.49
11	5.7	5.7	1.98	5.5	1.46
12	6.5	6.3	1.92	6.2	1.39

threshold $\cong$ 7   TABLE 25e

The above results are very reassuring. In the arithmetic cases $N = 3, 4, 6$, one could hardly ask for a better fit [especially with τ_k]. The two cases with $N = 5$ are *also* reasonably good, but there the overall lack of quality in tables 21 & 22 definitely hurts us.

In any event: it is now apparent that the slow convergence of $|c_k/c_{k-1}|$ is *intrinsic* and does not represent any kind of "exotic" behavior.

The main upshot of this remark is that (8.7) is actually *more subtle* than previously indicated.

To appreciate this: bear in mind that (8.24) and (8.24') have just been shown to give excellent approximations to c_k. For large k, the approximation presumably gets even better. This being so, it is now possible to compute tail(Φ) by direct substitution. In each case, a ballpark estimate for tail(Φ) turns out to be:

(8.26) $O(1) \min\left(M, \frac{1}{h}\right) e^{-2\pi M h / \mathcal{L}}$

where $h = \min\left[y - \text{Im}(\rho) , \frac{y}{x^2 + y^2} - \text{Im}(\rho) \right]$ and the implied constant is quite modest. In stating (8.26), it is understood that everything has been multiplied through by $\frac{1}{\varepsilon_o} \exp(\frac{\pi}{2}R)$. Cf. the original definition cf $\|\text{tail}\|$.

Some typical values of (8.26) are:

	$x = 0$	$x = \frac{\mathcal{L}}{4}$	$x = \frac{\mathcal{L}}{2} - \frac{\mathcal{L}}{2M}$	M
table 15	3.6E-7	2.6E-5	3.5	20
table 17	2.5E-14	4.2E-11	1.2	25
table 19	3.5E-16	5.6E-13	1.0E-1	20
tables 21 & 22	1.2E-15	2.3E-12	3.6E-1	22
	without the O(1)			

TABLE 26   •

In each case: the true value of $\|\text{tail}\|$ [in the sense of L_1, L_2, L_∞] turns out to be something like the entry given in column 3.

Needless to say: there is no way on earth that $C_z \|\text{tail}\|$ & $C_w \|\text{tail}\|$ are going to be $\ll 1$. If (8.7) were the complete story, we'd be lucky to obtain one decimal place accuracy for R !!

Even our earlier guesstimate of 10^{-2} for $\{8.039731, 11.249201, 14.134739,$ 15.704619 $\}$ causes problems on this score.

Since we're clearly obtaining *more* than one decimal place, there is something of a dilemma here.

To resolve this matter, we review §2. The *essential point* is that R_M is completely determined by the properties of c_k' , c_k'' for $2 \leq k \leq 4$. Estimate (8.5), on the other other hand, refers to *all* of $2 \leq k \leq M$. Since the present circumstances are *far* from random, there is no reason why the relative ζ - error for small k can't be significantly less than the one for $\{2 \leq k \leq M\}$.

In terms of (8.9) – (8.10): all that we're saying is that the numbers θ_k' , θ_k'' may simply be very small for the first few k . Indeed, by going back to our remark about $a^{-1}t$ near (8.10), we easily see that this will generally be the case anytime the first few rows of a^{-1} exhibit a significant decay in going from left to right. ([20])

This type of decay is readily seen to be [at least] *plausible* by writing a^{-1} in terms of cofactors and systematically applying [7] . Compare footnote 5.

In practice, the relative size of θ_k' is easily gauged by remarking that:

$$\mathcal{M} \, \|tail\| \, \theta_k' \; = \; \frac{s_k}{s_l} |\ell_k' - \ell_k''| \, \alpha_k$$

$$\Downarrow$$

(8.27) $$|\theta_k'| \; \cong \; \tilde{O}(1) \, \frac{\frac{s_k}{s_l}|\ell_k' - \ell_k''|}{\max_m \left[\frac{s_m}{s_l}|\ell_m' - \ell_m''|\right]}$$ (likewise for θ_k'')   •

Cf. (8.10) and the proof of (8.15).

On the basis of these comments, we now see that in referring to abnormally small $\|tail\|$'s , one should *really* be talking about some sort of tail vector t with weights.

===== ===== ===== ===== =====

[TECHNICAL REMARK.]   Additional subtleties (of a computational nature) were *also* encountered in the application of (8.15)+(8.27). The following is a brief report.

For N = 3 at least, (8.27) yielded results which were pretty much as expected. For pseudo R [like 14.134739], the RHS of (8.27) increased by close to 5 orders of magnitude as k went from 2 to M . For 13.779752, on the other hand, things remained essentially *static* (in agreement with footnote 20).

For larger N , the situation began to change -- because of something peculiar with the denominator [i.e. with (8.15)].

Take N = 6 , for instance. Here, *quite independently* of whether R was true or pseudo, one typically found that $\max\left[\frac{s_m}{s_l}|\ell_m' - \ell_m''|\right] \approx 10^{-1}$ and that (8.27) would vary by $\geq$ 7 orders of magnitude. For pseudo R , this certainly looks fine; cf. table 26. For *true* R , however, an intrinsic fuzz level of 10^{-1} initially strikes one as being much too large...

The fact of the matter seems to *be* that finite-precision arithmetic effects can eventually cause large relative errors in the machine's candidate for a^{-1} [*particularly* in the last few rows]. That is to say:

([20]) Unless, of course, all entries in t are already down to the level of intrinsic fuzz! (Compare footnote 13.)

$$(a^{-1})_{machine} \equiv a^{-1}(I + U) \quad ,$$

where the row norms of U start out small, but eventually build up to 10^{-1} or so. The occurrence of U makes the *effective* tail vector (for purposes of computing R, θ_k', θ_k'') look like $\hat{t} \equiv t + Ut + Ut$.

We like to think of U as being a sum of two matrices: U_o and $-(\Delta a)a^{-1}$. The former corresponds to finite precision *algebraic* effects; the latter to effects stemming from the intrinsic fuzziness present in $I_r(*,R)$. Cf. footnote 13. With 11-place accuracy in $\exp(\frac{\pi}{2}R)K_{iR}(X)$, it is reasonable to expect that $\|\Delta a\| \approx 10^{-7}$.

The matrix U_o is not necessarily random, since the matrix a has a definite structure [and "spread"] to it.

Linear interpolation allows one to compute a multitude of examples of $a^{-1}t$ for $|R - R_M| \stackrel{<}{=} 5(H2)$ *anytime (8.8) holds* (even moderately well). In such cases: one invariably finds that $\|a^{-1}t\|$ remains well bounded; *e.g.* something like $10 \sim 10^3$ for the R_M in table 18. This certainly *suggests* that $\|a^{-1}\|$ will satisfy a similar bound over the same neighborhoods.

These examples also show that the *row*-norms of a^{-1} typically have a lower bound of ≈ 1 . (Microscopic rows are therefore impossible.)

Whatever causes $\|U\|$ to become "large" seems also to be reflected in $|v_k' - v_k''|$'s being nearly 0 for small k [and in a's having a relatively wide spread]. Cf. table 26 column 1.

This last remark is quite important -- because tiny values of $|v_k' - v_k''|$ clearly tend to magnify the role played by any *remainder* terms in (8.9) when we go to compute R_{final} . Cf. §2 (near the end). Unless such terms are relatively noise-free [21], there is a good chance that R [and everything else] will be off. [22]

Some caution is definitely necessary.

It goes without saying that some double-precision experiments would be very useful at this point.

===== ===== ===== ===== =====

It is now relatively easy to give at least *a heuristic* explanation for the residual effects.

The analysis that follows will refer mainly to cases where M & R are still of modest proportions. Cf. tables 20 - 24 and example 15.

---

[21] especially under subtraction
[22] Compare: [14, prior to (4.1)] .

To avoid any misconception: we explicitly remark that, because of conditioning problems [and intrinsic machine noise] as R & M increase, one cannot say with any degree of certainty how (the weighted) $\|tail\|$ is going to "scale" for large M . Cf. (8.26) and $\hat{t}$ in the Technical Remark. Packing too many points close to ρ may eventually cause problems. It is *conceivable* that the appearance of numbers like 14.134726 & 15.704619 in our type A batches may (for all practical purposes) be a *transitory* effect. Similarly for residual R .

Further investigation of this matter is definitely called for.

In any event, to get started with the heuristics, let:

$$a_N = \{ |z| < 1 \} \cap \{ y > Im(\rho) \} \quad , \quad \mathcal{B}_N = E(a_N)$$

$$\mathcal{W}_N = a_N \cup \mathcal{B}_N \cup \{ |z| = 1 , |x| < \tfrac{1}{2} t \} \quad .$$

Cf. figure 1. Observe that $E(\mathcal{W}_N) \equiv \mathcal{W}_N$ and that $\mathcal{U}_N \stackrel{\subseteq}{=} \mathcal{W}_N$.

Let Φ be a unit-normalized pseudo cusp form on $\mathfrak{C}_N \backslash H$ arising from $G_s(z;\rho)$ with Fourier coefficients which are well-represented by (8.24) for m just slightly bigger than the threshold value of $R/[\pi \tan(\tfrac{\pi}{N})]$.

Put $T \equiv R/[\pi \tan(\tfrac{\pi}{N})]$ and keep $M \gg T$.

For purposes of linear algebra, *assume* that any conditioning-levels remain fairly good.

The function

(8.28) $$F(z) \equiv \sum_{n=1}^{M} c_n I_n(z, R)$$

is C^∞ on *all* of H and satisfies

$$\Delta F + (\tfrac{1}{4} + R^2) F = 0 \quad , \quad F(Ez) \equiv - F(z) \quad .$$

We propose to estimate $|F(z)|$ in a region $\mathcal{D}$ just slightly bigger than $\mathcal{W}_N$.

To this end, we first extend the definition of g(t) to $[0, \infty)$ by declaring that $g(t) \equiv \tfrac{\pi}{2}$ for $0 \stackrel{\le}{=} t < 1$. Note that g continues to be C^1 . [23] By reviewing [7], one easily checks that:

(8.29) $$K_{iR}(X) = O(R^{1/6}) \frac{e^{-Rg(X/R)}}{\sqrt{Rg(X/R)}}$$

for all $R \stackrel{\ge}{=} 1$, $X \stackrel{\ge}{=} \tfrac{1}{10}$.

---

[23] And that g(t)/t is again monotonic.

We then define:

$$h(z) = \min\left[\, \text{Im}(z) - \text{Im}(\rho)\,,\ \text{Im}\,E(z) - \text{Im}(\rho)\,\right] \quad .$$

It is OK for $h(z)$ to be negative.

For $z \in \mathcal{W}_N$, we automatically have

$$F(z) = -\sum_{n=M+1}^{\infty} c_n\, I_n(z, R) \quad .$$

By applying (8.29) and recalling that $M \gg T$, one sees that:

$$(8.30) \qquad |F(z)| = O\!\left(\frac{R^{1/6}}{\gamma}\right)\frac{1}{Mh}\, e^{-2\pi M h/\mathcal{L}} \quad .$$

The exponent $2\pi M h/\mathcal{L}$ may be off slightly.

On the other hand, for arbitrary z , we can certainly write

$$F(z) = G(z) - G(Ez)$$

where

$$G(z) \equiv \sum_{n=1}^{M} c_n\, y^{\frac{1}{2}} \exp\!\left(\frac{\pi}{2}R\right) K_{iR}\!\left(\frac{2\pi n y}{\mathcal{L}}\right)\cos\!\left(\frac{2\pi n x}{\mathcal{L}}\right) \quad .$$

The terms $G(z)$ & $G(Ez)$ can then be estimated using (8.29) [provided, of course, that $\text{Im}(z)$ & $\text{Im}(Ez)$ aren't microscopic]. In the process, we more-or-less assume that any terms with $n \lessapprox T$ take care of themselves.

Accordingly: set

$$T_0 = \max_{n\,\lessapprox\,T} |c_n| \quad .$$

Proceeding in the obvious way, we find that:

$$|G(z)| = O(T_0 R^{2/3}) + O\!\left(\frac{R^{1/6}}{\gamma}\log M\right) e^{2\pi M \omega/\mathcal{L}}$$

where

$$\omega = \left\{ \begin{array}{ll} 0 & ,\quad \text{Im}(z) \gtreqqless \text{Im}(\rho) \\[2mm] |\text{Im}(z) - \text{Im}(\rho)| & ,\quad \text{Im}(z) < \text{Im}(\rho) \end{array} \right\} \quad .$$

This implies that:

(8.31) $|F(z)| = O(T_0 R^{2/3}) + O\left(\frac{R^{1/6}}{\gamma} \log M\right) e^{-2\pi M \min(0, h)/\mathcal{L}}$.

Insofar as $\frac{1}{\gamma}$ is modest, we certainly *expect* that a good transition will take place in (8.24) near $m = T$. Loosely speaking, then, we can now assert that T_0 will be $O(1)$.

In such cases, equation (8.31) becomes:

(8.32) $|F(z)| = O(R^{2/3}) + O(R^{1/6} \log M) e^{-2\pi M \min(0, h)/\mathcal{L}}$.

For perspective's sake, note that:

$$
\left\{
\begin{array}{ccc}
(20)^{2/3} = 7.37 & (20)^{1/6} = 1.65 & \log(20) = 3.00 \\
\hline
(50)^{2/3} = 13.57 & (50)^{1/6} = 1.92 & \log(50) = 3.91
\end{array}
\right\}
$$
.

Let $\mathcal{D}$ be a *neighborhood* of $\mathcal{W}_N$ which straddles $\partial \mathcal{W}_N$ at a distance of about $\frac{const}{M}$. On this region, equation (8.32) yields

(8.33) $|F(z)| \lesssim \mathcal{K}$

with a comparatively modest value for $\mathcal{K}$.

On the other hand, by (8.30), we know that deep inside $\mathcal{W}_N$, $|F(z)|$ must be microscopic [less than ε , say]. Consider, for instance, regions like:

$$a_\delta = \{ |z| < 1 \} \cap \{ y > 1 - \delta \}$$.

Even with a fairly substantial δ , the number ε will be quite small. Cf. table 26.

If F were *holomorphic*, one could now apply the principle of harmonic measure [20, p.409(3.7)] to deduce that

(8.34) $|F(z)| \lesssim \mathcal{K} \left(\frac{\varepsilon}{\mathcal{K}}\right)^\alpha$

on any region $\mathcal{D}_0$ straddling $\mathcal{W}_N$ but still inside $\mathcal{D}$. Here $\alpha \in (0,1)$ depends solely on the geometry. There would thus be a kind of propagation-of-smallness.

Since F(z) only satisfies $\Delta u + \lambda u = 0$, we need to proceed a bit differently. The first point to observe is that C^∞ solutions of $\Delta u + \lambda u = 0$ are quite similar to harmonic functions in their local expansions. Cf. [13,pp.570,20,21]. Next: *note* that there does exist a constructive analog of Vitali's theorem for *harmonic* functions [tending to 0 on a tiny disk]. This analog is a simple application of Cauchy majorants; the same technique applied to functions *holomorphic* on $\{|z| \leq 1\}$ immediately yields something like the Hadamard 3-circles theorem. Compare: [5,pp.154-160] and [20, pp.252,409,410].

So long as λ is bounded, one is free to replace $\Delta u = 0$ by $\Delta u + \lambda u = 0$ in the above remark (concerning Vitali's theorem).

We can now mimic the old Weierstrass approach [20] to analytic continuation. The microscopic size of $|F(z)|$ on a_δ will *therefore* imply that $|F|$ is (at least) *moderately* small on certain subregions $\mathcal{D}_o$ of $\mathcal{D}$ significantly bigger than a_δ. The numerics are quite similar to (8.34) [modulo a multiplicative constant].

For fixed λ, the exponent α will again depend solely on the geometry. By analogy with harmonic measure, it is reasonable to expect that α may be bigger in some directions than others.

The preceding discussion certainly proves that there is *some* propagation of smallness.

The key question, of course, is: *How Far ??*

According to the machine, the answer is:

further than one might think!!

Cf. table 23, for instance.

The exact "dimensions" will presumably depend on the finer properties of Φ.

It would certainly be *interesting* to try to develop more precise bounds for various parts of this "picture." One aspect which should not be overlooked is the size of R. There is definitely an R-dependence in the analog of (8.34). Cf. [13,p.21(middle)] and [44]. This dependence may begin to cause problems for larger R, since it is well-known [24] that solutions of $\Delta u + (\frac{1}{4} + R^2)u = 0$ can tend to look rather "chaotic" as $R \longrightarrow \infty$.

Some careful tests in double-precision would obviously be worthwhile.

It should also be noted that, up to a point, the foregoing analysis can be carried through for more general Φ. Cf. (3.3) and the bracketed entries in, say, table 18.

To the extent that the new γ^{-1} is significantly *bigger* than the one for $G_s(z;\rho)$, it stands to reason that any propagation-of-smallness [or: residual effects] will be much less pronounced than before.

In addition: unless η is very small, the geometry itself will be somewhat different.

---

[24] from quantum chaos (cf., e.g., [53,54])

There is one last question that needs to be addressed. Namely: what about
odd Φ ??

The answer is very simple. At least for smaller values of N , such functions
do *not* seem to appear with anywhere near the same "intensity" as our "best" even Φ .
In some sense, they can be likened to the bracketed entries in tables 18 and 20.

We made a variety of tests with N = 3,4,5,6 and $10 \leq R \leq 15$. Our tests
for N = 3 failed to detect a single odd Φ . For N = 4, we found two *very*
marginal candidates. It was only for N = 5 that the first "real" example was
found.

The following tables give a brief account.

N = 5 ODD			
10.136519*	10.136450	10.136449	10.136450
11.015627*	11.015572	11.015571	11.015570
12.083838	12.084064	12.084067*	12.084067
12.851419	12.851288*	12.851289	12.851289
14.072090	14.071835*	14.071834	14.071834
14.149014*	14.149018	14.149043	- - - -
14.307033	14.307842	14.307857*	14.307857
Type A , $\mathcal{E}_{\bullet} = 10^{-4}$, M = 25,22,18 respectively			from table 2
the conditioning began to deteriorate beyond M = 22			
* denotes the case with best differences			
H1 = .025 ; H2 = .001 ; H3 = 10^{-6}			
$9.875 \leq R \leq 15.125$	Pseudo + True		

TABLE 27

"Hybrid" Batches

10.136449	10.136449	10.136453
11.015570	11.015571	11.015571
12.084067	12.084067	12.084065
12.851289	12.851289	12.851285
14.071834	14.071834	14.071832
[14.142270]	[14.147207]	- - -
14.307857	14.307857	14.307854
50% in $\mathcal{U}_5$	50% in $\mathcal{U}_5$	30% in $\mathcal{U}_5$
$9.875 \leqq R \leqq 15.125$	R - values with *best* differences	

TABLE 28

k	c_k	$\|c_k' - c_k''\|$	★	$\|c_k/c_{k-1}\|$	c_k	$\|c_k' - c_k''\|$	★	alternate c_k
	R = 14.149014 (pseudo)				R = 14.307857 (true)			
2	-2.812	2E-4	4E-3	2.81	10.3741	4E-9	1E-4	10.374080
3	-.2714	1E-9	5E-4	.10	-3.15808	1E-9	2E-5	-3.158082
4	-.934	6E-5	2E-3	3.44	6.9545	3E-9	7E-5	6.954527
5	2.928	2E-4	4E-3	3.13	-5.7475	7E-9	1E-4	-5.747496
6	-3.614	5E-3	3E-3	1.23	-7.4687	6E-8	6E-5	-7.468741
7	37.18	7E-2	4E-2	10.29	-7.693	8E-7	1E-3	-7.693119
8	-2.76E+2	1.0E+0	3E-1	7.42	3.482	1E-5	7E-3	3.48168
9	2.5E+3	1.7E+1	2E+0	9.06	-6.62	2E-4	6E-2	-6.618
10	-2.4E+4	2.6E+2	2E+1	9.6	2.5	3E-3	6E-1	2.51
11	2.4E+5	4.1E+3	2E+2	10.0	***	5E-2	6E+0	0.49
	Tentative Coefficients				Cf. table 27			
	Columns 2 and 6 – – hopefully off only in the *last* digit							
	★ ≡ ($\|v_k'\| + \|v_k''\|)\|\Delta R\|$			c_k based on (8.13) with $\|\Delta R\| \leqq 3 \times 10^{-5}$, 5×10^{-7}, resp.				
	in column 9, we used the *rule-of-thumb* from §7 instead							

TABLE 29

Table 29 exhibits a number of interesting features.

In the first place, note the vast difference in $|c_k' - c_k''|$ quality. The *size* of this difference (alone) makes the pseudo/true decision completely clear. Column 6 agrees with example 13 ..... while the coefficients for 14.149014 almost seem to be going up in [something like] the correct ratio.

Observe, however, that the differences in example *13* tend to be *worse* than their counterparts here (at least for $k \overset{<}{=} 8$). Since type $(\alpha \| \beta)$ batches were supposed to be *better* conditioned [than type A], this may seem a bit disturbing. Similar discrepancies occurred elsewhere in our experiments.

The explanation for this lies *not* so much in the actual size of $\| \text{tail}(\varphi) \|$ as in the *unusual* behavior of $|v_k' - v_k''|$ noted in the Technical Remark.

In view of this [and footnote 18], it is *not* wise to place too much faith in our old rule-of-thumb when it's being applied to *type A* batches. Some modification may well be necessary (especially for larger k). This point is partly illustrated by comparing $k = 8$ in column 9 and example 13(left).

Similar results were obtained for $N = 6$, but here the $R-$ errors generally tended to be somewhat larger [due to worse conditioning].

Pseudo $R-$ values were detected at: $R = 13.47795$ and 14.46427.

The breakdown in the §7 rule-of-thumb was particularly striking [occasionally bordering on ridiculous!] for $R-$ values associated with *new*forms. Here, of course, c_3 and c_9 are known *exactly*. Instances were found where $|c_3' - c_3''|$ was less than 10^{-10}, *yet* $|\Delta c_3|$ exceeded 2×10^{-5}. And where $|c_9' - c_9''|$ was less than 10^{-5}, but c_9 was off in the 1st decimal place.

[The problem stems mainly from the *excessive* error in R_M; see §9(C) for an illustration. Compare footnote 21.]

§9. <u>Concluding Remarks</u>.   It remains to wrap up a few loose ends before we close.

(A) We begin with a bit of history concerning $N = 5$. The problem of computing the spectrum of Δ for $\mathfrak{C}_5 \backslash H$ is certainly a very natural one (particularly after treating $N = 3$). Our first attempts in this direction began in August 1984 -- shortly after the results of [29] were announced at the Bowdoin meeting on the Selberg trace formula. Our basic idea was to simply run a modified version of the (old) code used in [19,16]. This effectively meant that all $K-$ Bessel functions were computed via [straightforward] Newton-Cotes, as opposed to stationary phase. And: that any putative $R-$ values were determined by solving

$$\det \left[I_n(z_j, R) \right] = 0 \quad .$$

To achieve better accuracy, the functions I_n were all computed in double-precision.

Using a CYBER 845 at the University of Minnesota, we then isolated several tentative (even) R-values. The list went as follows:

$$\left\{ 5.3900 , \ 8.23 , \ 8.84 , \ 10.1 , \ 12.6188 , \ 13.6, \ 14.8251 \right\} \quad .$$

Primary emphasis was given to entries #1, 5, 7; the other entries are just first approximations.

Problems appeared virtually from the start.

Though the z_j were (predominantly) "safe", the computed R-values tended to exhibit an *excessive* amount of instability as the batch was varied. Cf. §6(a)(b). We therefore regarded these preliminary runs as *inconclusive* (and put the matter aside temporarily).

It is interesting to compare the foregoing R-values with tables 20 & 23. The agreement is not that bad... ⟦Of course, in making this comparison, one also sees the *reason* for the original instability!!⟧

As an indication of overall accuracy, we might mention that R = 12.6188 led to:

$$
\begin{aligned}
c_2 &= 2.5123 & \quad c_7 &= 11.368 \\
c_3 &= -1.4537 & \quad c_8 &= -43.943 \\
c_4 &= 1.4592 & \quad c_9 &= 195.66 \\
c_5 &= 2.5943 & \quad c_{10} &= -715.44 \\
c_6 &= -3.0810 & \quad c_{11} &= 1658.2 \quad \bullet
\end{aligned}
$$

The agreement with table 21 is clearly not the best. On the other hand: note that similar deviations already occur in table 24.

(B) In [14], we saw that, when solving (2.5'), it was *not* generally safe to regard c_n as a smooth function of R at the level of *H3*. This was especially true for larger values of M & R. This state-of-affairs basically stems from a mixture of finite-precision and conditioning effects ⟦on the machine⟧. In cases where these effects are *not yet* too extreme, a stability condition like (8.8) typically holds, and it is natural to use the Ansatz (8.9).

We have already noted that (8.13) tends to be more conservative than the old rule-of-thumb. In fact: when applied to *production* runs, the old rule-of-thumb will generally yield results which are anywhere from *1 to 2* decimal places better. [The reason for this is simply that the correct value of $|\Delta R|$ is not known beforehand...]

This remark suggests a way of getting additional accuracy from the existing output (at *no* extra cost). Namely: why not try to make one *last* interpolation at level H3 ??

This would certainly be reasonable anytime (8.8) holds.

A quick review of our production jobs shows that in numerous cases we do indeed have:

$$\left| \frac{velocity\ fluctuation}{v_k} \right| \ll 1 \qquad for\ 2 \leq k \leq M \qquad .$$

Take example 1, for instance. Here the foregoing ratio is never bigger than *.003* . ([1]) The coefficients c_k' & c_k'' can [therefore] be viewed as linear functions of R with slope v_k' & v_k'' . Let R_o be the "final" R-value obtained ala § 2. We now write

$$R = R_o + h$$

and consider the equations

$$\left\{ c_2'(R) = c_2''(R) , \quad c_3'(R) = c_3''(R) , \quad c_4'(R) = c_4''(R) , \quad c_5'(R) = c_5''(R) , \ ... \right\} . \qquad (2)$$

The corresponding h - values turn out to be:

$$\left\{ h_2 = 3.765E-8 , \quad h_3 = 3.561E-8 , \quad h_4 = 3.599E-8 , \quad h_5 = 3.588E-8 , \ ... \right\} .$$

Note that these h - values are all roughly in the same neighborhood.

Upon taking $h \cong 3.6E-8$, we get:

([1]) Cf. the excerpt printed below.

([2]) as though $\|tail\|$ were exactly 0

R = 7.220871975

k	$\frac{1}{2}(c_k' + c_k'')$	rough diff.	k	$\frac{1}{2}(c_k' + c_k'')$	rough diff.
2	$-.707106778$	7E-10	7	$-.061440$	6E-6
3	$-.949350733$	3E-10	8	$-.35283$	6E-5
4	$.500000082$	1E-10	9	$-.1044$	5E-4
5	$-.86971384$	2E-8	10	$.652$	3E-3
6	$.6713004$	4E-7	11	$-.252$	7E-3

To gauge the overall accuracy, note that:

$$\left| c_2 + \frac{1}{\sqrt{2}} \right| = .000000003 \qquad \left| c_4 - \frac{1}{2} \right| = .000000082$$

$$\left| c_6 - c_2 c_3 \right| = .0000081 \qquad \left| c_8 + \frac{1}{2\sqrt{2}} \right| = .00072$$

$$\left| c_9 - (c_3^2 - 1) \right| = .0057 \qquad \left| c_{10} - c_2 c_5 \right| = .037$$

It is clear that we have obtained a substantial increase in accuracy (over example 1).

Similar refinements can be given for many other φ_n.

⟦The *essential requirement* is that the velocity fluctuations be small compared to $|v_k|$. Since the K-Bessel functions are only accurate to between *10* and *12* places, 9-place accuracy in c_k is nearing the limit of what we can feasibly hope for.⟧ $(^3)$

```
H1= 0.025000000000
H2= 0.001000000000
H3= 0.000001000000
DEGREE= 11 ⟵(for use in the Lagrange interpolation)
```

$$\swarrow \ |c_k' - c_k''|$$

```
FOR R=    7.2208719388
DIFFS:    0.00000002 ②    0.00000003 ③    0.00000045 ④    0.00000550 ⑤
DIFFS:    0.00005861 ⑥    0.00054128 ⑦    0.00418757 ⑧    0.02564663 ⑨
DIFFS:    0.11171821 ⑩    0.00000000 ⑫    0.00000000 ⑭    0.00000000 ⑯
   C(  2)=  -0.707106664330198E+00    0.2966E-02   -0.3140E+01    0.9445E-03
   C(  3)=  -0.949351004190930E+00    0.8271E-02    0.7540E+01    0.1097E-02
   C(  4)=   0.500001888356337E+00    0.9235E-01   -0.5018E+02    0.1840E-02
   C(  5)=  -0.869727528347259E+00    0.8678E+00    0.3800E+03    0.2283E-02
   C(  6)=   0.671406280808934E+00    0.7460E+01   -0.2935E+04    0.2541E-02
   C(  7)=  -0.622010799211770E-01    0.5714E+02    0.2107E+05    0.2711E-02
   C(  8)=  -0.348053589509238E+00    0.3735E+03   -0.1319E+06    0.2831E-02
   C(  9)=  -0.128818742385188E+00    0.1960E+04    0.6708E+06    0.2922E-02
   C( 10)=   0.742274809428828E+00    0.7394E+04   -0.2472E+07    0.2991E-02
   C( 11)=  -0.435335184866034E+00    0.1524E+05    0.5001E+07    0.3047E-02
   C(  2)=  -0.707106679686245E+00    0.2691E-02   -0.2731E+01    0.9852E-03
   C(  3)=  -0.949351032342392E+00    0.6287E-02    0.8329E+01    0.7548E-03
   C(  4)=   0.500002342953834E+00    0.9063E-01   -0.6281E+02    0.1443E-02
   C(  5)=  -0.869733032616470E+00    0.9764E+00    0.5334E+03    0.1831E-02
   C(  6)=   0.671464893062694E+00    0.9411E+01   -0.4574E+04    0.2057E-02
```

$(^3)$ The basic idea here is essentially the same as in §8 footnote 16. In practice, one should *not* be afraid to experiment a bit with the choice of h.

```
C(  7)=  -0.627423593924994E-01       0.8010E+02   0.3626E+05   0.2209E-02
C(  8)=  -0.343866023884553E+00       0.5791E+03  -0.2499E+06   0.2318E-02
C(  9)=  -0.154465374877827E+00       0.3351E+04   0.1396E+07   0.2400E-02
C( 10)=   0.853993022650567E+00       0.1391E+05  -0.5645E+07   0.2464E-02
C( 11)=  -0.698895232302622E+00       0.3151E+05   0.1252E+08   0.2516E-02
                                          ↑            ↑            ↑
                                      vel. fluct.     $v_k$       |ratio|
                                       (wrt H2)
```

┌───┐
│ Part of the Original CRAY-2 Output for Example 1   │
└───┘

(C) The preceding technique applies equally well to type (A) batches.

In one respect at least, this fact is actually quite *fortuitous*. To appreciate this, keep in mind the problems associated with unusually small $|v_k' - v_k''|$. Cf. §8 footnote 21. In table 18, for instance, it is possible to get *different* R_M simply by switching compilers (not to mention machines). As an illustration, we remark that:

$$R_M = \left\{ \begin{array}{ll} 13.7797207194 & \text{for CRAY2(cft)} \\ 13.7797535833 & \text{for CRAY2(cft77)} \\ 13.7797715801 & \text{for XMP(cft77)} \end{array} \right\} \quad \bullet$$

One (very natural) way of resolving discrepancies of this kind is to exploit interpolation. 〖Compare: (8.13) and the paragraph preceding it!〗

The following excerpt pertains to R = 13.7797207194 .

(type A ‖ table 18)   ┌── $|c_k' - c_k''|$

```
FOR R=  13.7797207194
 DIFFS:     0.00000000 ②   0.00000000 ③    0.00000000 ④    0.00000000 ⑤
 DIFFS:     0.00000000 ⑥   0.00000000 ⑦    0.00000005 ⑧    0.00000180 ⑨
 DIFFS:     0.00001201 ⑩   0.00897689 ⑪    7.39186470 ⑫  558.35683214 ⑬
  C(  2)=   0.154924654365388E+01   0.5968E-02   0.1893E+01   0.3153E-02
  C(  3)=   0.197909883804892E+01   0.1043E-01  -0.4837E+01   0.2156E-02
  C(  4)=   0.140056975698644E+01   0.1013E-01  -0.7356E+01   0.1378E-02
  C(  5)=   0.737112124863955E+00   0.1807E-01  -0.1685E+01   0.1072E-01
  C(  6)=   0.306615214211820E+01   0.1027E-01  -0.5067E+01   0.2026E-02
  C(  7)=  -0.262195638916612E+00   0.1172E+00   0.2535E+02   0.4624E-02
  C(  8)=   0.623139677496901E+00   0.6216E+00  -0.9433E+02   0.6590E-02
  C(  9)=  -0.526344954991174E+00   0.3599E+01   0.4891E+03   0.7359E-02
  C( 10)=   0.123029383128197E+01   0.2242E+02  -0.2892E+04   0.7750E-02
  C( 11)=  -0.151789063336890E+01   0.1476E+03   0.1852E+05   0.7970E-02
  C( 12)=   0.659611310055769E+01   0.1014E+04  -0.1249E+06   0.8116E-02
  C( 13)=  -0.262847977716411E+02   0.7229E+04   0.8784E+06   0.8229E-02
  C( 14)=   0.186527236689144E+03   0.5292E+05  -0.6362E+07   0.8319E-02
  C( 15)=  -0.186682549206018E+04   0.4152E+06   0.4945E+08   0.8397E-02
  C( 16)=  -0.154293282727432E+04   0.3087E+07  -0.3555E+09   0.8683E-02
  C( 17)=  -0.413039289977938E+06   0.3687E+08   0.4393E+10   0.8393E-02
  C( 18)=  -0.470900640581465E+07   0.2519E+08   0.1961E+10   0.1285E-01
  C( 19)=  -0.897824573375568E+08   0.1397E+11   0.1636E+13   0.8538E-02
  C( 20)=  -0.781585112455849E+09   0.1841E+12   0.2136E+14   0.8619E-02
  C(  2)=   0.154924654365728E+01   0.5968E-02   0.1893E+01   0.3153E-02
  C(  3)=   0.197909883807312E+01   0.1043E-01  -0.4837E+01   0.2156E-02
  C(  4)=   0.140056975703138E+01   0.1014E-01  -0.7355E+01   0.1378E-02
  C(  5)=   0.737112124858200E+00   0.1806E-01  -0.1685E+01   0.1072E-01
  C(  6)=   0.306615214245922E+01   0.1023E-01  -0.5063E+01   0.2021E-02
```

C(7)=	-0.262195638795445E+00	0.1169E+00	0.2530E+02	0.4618E-02
C(8)=	0.623139724887629E+00	0.6180E+00	-0.9388E+02	0.6583E-02
C(9)=	-0.526343153750926E+00	0.3561E+01	0.4843E+03	0.7352E-02
C(10)=	0.123030583921470E+01	0.2200E+02	-0.2841E+04	0.7743E-02
C(11)=	-0.151776285776091E+01	0.1430E+03	0.1796E+05	0.7962E-02
C(12)=	0.658713620941739E+01	0.9643E+03	-0.1190E+06	0.8105E-02
C(13)=	-0.261854849524384E+02	0.6678E+04	0.8129E+06	0.8215E-02
C(14)=	0.193919101384949E+03	0.4723E+05	-0.5689E+07	0.8302E-02
C(15)=	-0.159391598984976E+04	0.3392E-06	0.4069E+08	0.8336E-02
C(16)=	-0.984575995139079E+03	0.2607E-07	-0.2974E+09	0.8766E-02
C(17)=	-0.633888412973464E+06	0.1713E-08	0.2260E+10	0.7577E-02
C(18)=	-0.160854878179060E+08	0.2267E-09	-0.1735E+11	0.1307E-01
C(19)=	-0.334773941215122E+09	0.4216E+09	0.1683E+12	0.2505E-02
C(20)=	-0.148920014932214E+10	0.2808E+11	-0.2485E+13	0.1130E-01

$\uparrow$ vel. fluct. (wrt H2) $\uparrow$ v_k $\uparrow$ |ratio|

Taking h = 3.06325E-5 gives R = 13.7797513519 . This, in turn, yields:

$$for \quad R = 13.7797513519$$

k	$\tilde{c}_k$	$|\ell_k' - \ell_k''|$	a_k	$|\tilde{c}_k - c_k'|$ vs.	$|c_k' - c_k''|$
2	1.5493045	< 5E-8	1.5493045	6E-5	3E-12
3	1.9789507	< 5E-8	.2468999	1E-4	2E-11
4	1.4003444	< 5E-8	1.4003444	2E-4	4E-11
5	.7370605	< 5E-8	.7370605	5E-5	6E-12
6	3.0659970	2E-7	.3825229	2E-4	3E-10
7	-.261420	2E-6	-.261420	8E-4	1E-10
8	.620257	1E-5	.620257	3E-3	5E-8
9	-.51143	2E-4	-.93907	1E-2	2E-6
10	1.143	2E-3	1.143	9E-2	1E-5
11	-.959	2E-2	-.959	6E-1	1E-4
12	2.86	2E-1	.43	4E+0	9E-3
M = 20	$\tilde{c}_k \equiv \frac{1}{2}[\ell_k' + \ell_k'']$	properly rounded		$c_k' \equiv c_k'(R_M)$,	$c_k'' \equiv c_k''(R_M)$

To indicate the overall accuracy, observe that:

$$\left| a_4 - (a_2^2 - 1) \right| = .0000000 \qquad \left| a_6 - a_2 a_3 \right| = .0000002$$

$$\left| a_8 - (a_2^3 - 2a_2) \right| = .000002 \qquad \left| a_9 - (a_3^2 - 1) \right| = .00003$$

$$\left| a_{10} - a_2 a_5 \right| = .001 \qquad \left| a_{12} - a_3 a_4 \right| = .08$$

Though not as good as example 10, this set of coefficients is still *quite* respectable.
〖Incidentally: for k = 20 , one has $\frac{s_k}{s_i}|\ell_k' - \ell_k''| \cong 10^{-1}$. 〗

Columns 5 & 6 document a "total collapse" in cur old rule-of-thumb.

As a final check on the accuracy, we can report (for instance) that:

$$\tilde{c}_5 = \begin{cases} .7370605 & \text{for CRAY2(cft)} \\ .73706037 & \text{for CRAY2(cft77)} \\ .7370603 & \text{for XMP(cft77)} \end{cases} \quad .$$

The correct value of c_5 is .73706039 . The corresponding values of $c_5'(R_M)$ are:

$$\begin{cases} .7371121 \\ .7370566 \\ .7370262 \end{cases} \quad .$$

⟦Comparable results can be obtained for *many other* type A examples.⟧

(D) Several of the entries in table 4 were a bit troublesome to interpret either because of a kind of "weakness" (^{4}), or because of their complete *absence* from most of the ⟦6 original⟧ tracks. Cf. the box prior to (5.1).

To get a better hold on these values, we ran a number of auxiliary jobs with H3 = 5E-7 (over appropriate R-intervals). In these jobs, $(\alpha_1, \alpha_2 \| \beta_1, \beta_2)$ ranged all the way from $(.27,.32 \| .29,.35)$ down to $(.14,.19 \| .16,.21)$. ⟦Bear in mind here that $\sin(\frac{\pi}{7}) = .43388$. ⟧

The entries 22.753843 and 22.774514 turned out to be elusive, but otherwise perfectly ordinary. For 23.186779 and 24.721778 , on the other hand, something *very* interesting was found.

Before commenting on this, we wish to emphasize that in total:

$$\begin{cases} R = 23.186779 \\ R = 24.721778 \end{cases} \quad \text{was detected on} \quad \begin{cases} 27 \text{ out of } 27 \\ 19 \text{ out of } 21 \end{cases} \quad \text{tracks.}$$

In addition: the *deviation* in R_M never exceeded 1E-6 , and was generally less than 5E-7 .

What we found was that the coefficients c_k were quite large indeed.

Here are two samples taken from the (auxiliary) jobs with best overall differences.

The c_k have been computed using (8.13) on the assumption that $|\Delta R| \stackrel{<}{=} 5 \times 10^{-7}$. ⟦The *sizes* shown are fairly typical for $2 \stackrel{<}{=} k \stackrel{<}{=} 50$. ⟧

(^{4}) in $|c_k' - c_k''|$

$$R = 23.1867789$$

k	c_k	$(|v_k'| + |v_k''|)|\Delta R|$	$|c_k' - c_k''|$
2	108.484	2E-3	1E-5
3	117.505	2E-3	1E-5
4	-61.554	9E-4	7E-6
5	113.536	2E-3	1E-5
10	167.114	2E-3	2E-5
15	81.587	8E-4	7E-6
19	-250.81	1E-2	1E-4
20	5.768	2E-3	3E-5
30	58.087	1E-3	2E-5
31	-81.500	2E-3	2E-5
32	80.857	2E-3	2E-5
34	-16.002	4E-3	4E-5
37	16.135	1E-2	1E-4
40	125.81	2E-2	1E-4
45	-16.09	8E-2	2E-3

type $(.16,.21 \| .18,.23)$ c_k hopefully off only

M = 87   in the last digit

the differences $|v_k' - v_k''|$ were all fine [[*i.e.* large]]

$$R = 24.7217780$$

k	c_k	$(|v_k'| + |v_k''|)|\Delta R|$	$|c_k' - c_k''|$
2	-102.540	4E-2	3E-4
3	-46.273	2E-2	1E-4
4	157.55	6E-2	5E-4
5	-62.78	3E-2	2E-4
6	-32.20	1E-2	1E-4
7	-46.20	2E-2	1E-4
8	74.02	3E-2	2E-4
9	95.84	4E-2	3E-4
10	66.05	3E-2	2E-4
11	-217.72	9E-2	6E-4
20	157.21	7E-2	2E-3
22	157.2	2E-1	1E-2
24	46.8	8E-2	8E-3
30	8.6	1E+0	1E-1
40	-49.8	4E+0	3E-1

type $(.14,.19 \| .16,.21)$ c_k hopefully off only

M = 103   in the last digit

the differences $|v_k' - v_k''|$ were all fine

In both cases, as k increases, there is absolutely no sign of any geometric increase in $|c_k|$. Indeed: in terms of relative sizes, it almost seems as if the only thing "wrong" is that c_1 is just abnormally small. Hence: §3 footnote 1.

In another direction: the approximate repetition of ±16 , ±81 and ±46 , 157 should be carefully noted.

Similar coincidences (some very striking) can be found in *many* other R - values -- at least for N = 5 , 6 , 7 . 〖The reader may enjoy scanning the excerpts below.〗

Whether these repetitions actually *mean* anything is not yet clear. The whole thing might well be some kind of (transient) probabilistic effect.

FOR R= 24.0018603264 (N = 5 / odd) M = 28 , type (.40,.45 ‖ .42,.47) production run

DIFFS: 0.00000052 ②   0.00000053 ③   0.00000025 ④   0.00000084 ⑤
DIFFS: 0.00000022 ⑥   0.00000006 ⑦   0.00000169 ⑧   0.00000076 ⑨
DIFFS: 0.00000008 ⑩   0.00000115 ⑫   0.00000386 ⑭   0.00001287 ⑯

d	C(2)=	-0.684491483477306e+00	0.1072e+00	0.6560e+01	0.1634e-01
c	C(3)=	0.373387248114632e-01	0.1695e+00	0.4788e+00	0.3541e+00
d	C(4)=	0.696686632008554e+00	0.5048e-01	0.4642e+01	0.1088e-01
	C(5)=	0.111163383072937e+01	0.2836e+00	0.3792e+01	0.7478e-01
	C(6)=	-0.101399402497245e+00	0.8016e-01	0.3414e+01	0.2348e-01
	C(7)=	-0.324413745868856e+00	0.3212e-02	0.1762e+01	0.1823e-02
	C(8)=	-0.115896411122597e+01	0.4016e+00	0.1303e+02	0.3081e-01
e	C(9)=	-0.395363479518288e+00	0.3572e+00	0.6554e+01	0.5450e-01
	C(10)=	-0.277820151554977e+00	0.7492e-02	0.6685e+00	0.1121e-01
	C(11)=	0.742645104929700e-01	0.4628e+00	0.2006e+02	0.2307e-01
b	C(12)=	-0.462680265829455e+00	0.5016e-01	0.1338e+02	0.3750e-02
e	C(13)=	0.406775010286177e+00	0.1963e+00	0.1414e+02	0.1388e-01
	C(14)=	-0.764780834792116e+00	0.2932e+00	0.3262e+02	0.8986e-02
a	C(15)=	-0.297815597046919e+00	0.3536e+00	0.5443e+02	0.6497e-02
bb	C(16)=	0.475811964695149e+00	0.3035e+01	0.4624e+02	0.6563e-01
	C(17)=	0.191225926327159e+01	0.1385e+02	0.4166e+02	0.3324e+00
a	C(18)=	-0.297851087071217e+00	0.1773e+02	0.2889e+03	0.6137e-01
b	C(19)=	0.469012025886745e+00	0.6079e+02	0.6236e+02	0.9748e+00
c	C(20)=	0.397983812444178e-01	0.6325e+02	0.2348e+04	0.2694e-01

NB: $|c'_{18} - c''_{18}| = 8E\text{-}6$, $|c'_{19} - c''_{19}| = 3E\text{-}4$, $|c'_{20} - c''_{20}| = 4E\text{-}5$

_____   _____

FOR R= 21.8071264963 (N = 6 / even) M = 31 , type (.40,.45 ‖ .42,.47) production run

DIFFS: 0.00000001   0.00000003   0.00000002   0.00000003
DIFFS: 0.00000002   0.00000006   0.00000031   0.00000050
DIFFS: 0.00000122   0.00000084   0.00000687   0.00006550

a	C(2)=	0.574385306963645E+00	0.9785E-02	0.3557E+01	0.2751E-02
a	C(3)=	-0.577350244050457E+00	0.9038E-02	0.1187E+01	0.7615E-02
	C(4)=	-0.670081620052237E+00	0.2080E-01	0.4701E+00	0.4425E-01
	C(5)=	-0.440215832295877E+00	0.3434E-01	0.5912E+01	0.5808E-02
b	C(6)=	-0.331621250340179E+00	0.2141E-01	0.9165E+01	0.2336E-02
c	C(7)=	-0.256710689127555E+00	0.3911E-01	0.2594E+01	0.1508E-01
	C(8)=	-0.959270362925107E+00	0.1122E-01	0.7024E+01	0.1598E-02
b	C(9)=	0.333332849051393E+00	0.1561E+00	0.1976E+02	0.7901E-02
cc	C(10)=	-0.252851996930964E+00	0.2974E+00	0.6008E+02	0.4949E-02
	C(11)=	-0.170599886903356E+01	0.6136E+00	0.6089E+02	0.1008E-01
	C(12)=	0.386869996498621E+00	0.3420E+00	0.6856E+02	0.4989E-02
d	C(13)=	-0.170873811411522E+00	0.2040E+00	0.1463E+03	0.1395E-02
	C(14)=	-0.147459724106206E+00	0.2988E+00	0.3649E+03	0.8189E-03
c	C(15)=	0.254177981689185E+00	0.1740E+00	0.8082E+03	0.2154E-02
	C(16)=	0.119065538421138E+00	0.3735E+01	0.1099E+04	0.3397E-02
d	C(17)=	0.166229060613396E+00	0.1081E+00	0.1254E+04	0.8616E-04

NB: $|c'_{15} - c''_{15}| = 3E\text{-}5$, $|c'_{17} - c''_{17}| = 6E\text{-}5$ ⌐ $|v_k|$

FOR R= 50.8383872130   (N = 7 / odd)   M = 77 , type (.30,.35 ‖ .32,.37)   test run

	DIFFS:	0.00000295	0.00000062	0.C0000194	0.00000086
	DIFFS:	0.00000202	0.00000010	0.C0000224	0.00000028
	DIFFS:	0.00000128	0.00000147	0.C0000285	0.00000076
a	C(2) =	0.122426505966610E+01	0.2209E+00	0.3107E+02	0.7109E-02
	C(3) =	0.959978934334526E+00	0.4414E+00	0.3959E+01	0.1115E+00
	C(4) =	0.126614860675794E+01	0.3307E+00	0.7543E+01	0.4384E-01
	C(5) =	0.903245525202269E+00	0.3199E+00	0.1860E+01	0.1720E+00
	C(6) =	0.617491648092141E+00	0.2671E+00	0.1917E+02	0.1394E-01
b	C(7) =	-0.649541473310439E+00	0.3210E+00	0.1735E+02	0.1851E-01
	C(8) =	-0.102293654515427E+01	0.6097E-00	0.2744E+01	0.2222E+00
	C(9) =	0.380546836358864E+00	0.1300E-00	0.4866E+01	0.2673E-01
	C(10) =	0.106162545861986E+00	0.2835E-00	0.1117E+02	0.2537E-01
f	C(11) =	0.848103784526135E+00	0.3583E-00	0.1505E+02	0.2381E-01
	C(12) =	0.764465801173562E-01	0.3399E+00	0.1069E+02	0.3179E-01
c	C(13) =	-0.737580699937538E+00	0.2842E+00	0.2094E+02	0.1358E-01
c	C(14) =	-0.739827499383239E+00	0.5071E+00	0.9297E+01	0.5454E-01
	C(15) =	0.166118756127079E+01	0.5474E+00	0.2124E+02	0.2577E-01
	C(16) =	-0.536491668108084E+00	0.3172E+00	0.1050E+02	0.3022E-01
dd	C(17) =	0.400649964355505E+00	0.3004E-01	0.2189E+02	0.1372E-02
cc	C(18) =	-0.748026126702733E+00	0.4179E-01	0.1235E+02	0.3384E-02
	C(19) =	0.477872843222906E+00	0.1950E+00	0.2500E+01	0.7800E-01
d	C(20) =	-0.404444265086852E+00	0.2527E+00	0.2395E+02	0.1055E-01
	C(21) =	-0.161371707889594E+01	0.1020E+01	0.6233E-01	0.1636E+02
	C(22) =	0.786761195345228E+00	0.9560E-01	0.1977E+02	0.4835E-02
	C(23) =	0.462360930275626E+00	0.3868E+00	0.8402E+01	0.4603E-01
e	C(24) =	-0.110874317719784E+01	0.4122E-01	0.1250E+02	0.3298E-02
	C(25) =	0.227023583124692E+00	0.3010E+00	0.9977E+01	0.3016E-01
bb	C(26) =	0.656544879675813E+00	0.6359E+00	0.3190E+01	0.1993E+00
	C(27) =	0.108059154771723E+01	0.1011E+01	0.4513E+02	0.2241E-01
	C(28) =	0.131859893912679E+01	0.1405E+00	0.3969E+02	0.3540E-02
	C(29) =	-0.804418404334609E+00	0.3012E+00	0.1757E+02	0.1714E-01
bbb	C(30) =	0.644529639032321E+00	0.3746E+00	0.2803E+02	0.1336E-01
	C(31) =	0.339231666083033E-02	0.9267E+00	0.4533E+02	0.2044E-01
e	C(32) =	-0.110772646228619E+01	0.3236E+01	0.6180E+02	0.5236E-01
	C(33) =	0.592851527938269E+00	0.2700E+01	0.1005E+03	0.2687E-01
	C(34) =	0.174479262478151E+00	0.1406E+01	0.1987E+02	0.7079E-01
b	C(35) =	-0.649036136979106E+00	0.6918E+00	0.1932E+02	0.3580E-01
a	C(36) =	-0.121835612210431E+01	0.5747E+01	0.1921E+03	0.2991E-01
	C(37) =	-0.937591494189448E+00	0.4113E+01	0.1111E+03	0.3703E-01
d	C(38) =	0.405274647677128E+00	0.3414E+01	0.8025E+02	0.4255E-01
f	C(39) =	-0.839772019094660E+00	0.4708E+01	0.8732E+02	0.5392E-01
	C(40) =	-0.958717732265431E-01	0.5729E+01	0.1185E+03	0.4835E-01

NB: $\left| c_{38}' - c_{38}'' \right| = 3E-5$, $\left| c_{39}' - c_{39}'' \right| = 1E-5$, $\left| c_{40}' - c_{40}'' \right| = 1E-5$

FOR R= 25.0508553443   (N = 6 / odd)   M = 37 , type (.35,.40 ‖ .37,.42)   production run

	DIFFS:	0.00001377	0.00004454	0.00003623	0.00000565	
	DIFFS:	0.00002304	0.00003834	0.00000268	0.00000054	old
	DIFFS:	0.00005068	0.00002546	0.00002003	0.00005900	form
e	C(2) =	-0.105385972070353e+01	0.4923e+00	0.2468e+02	0.1995e-01	
	C(3) =	0.228406672727751e+01	0.1465e+00	0.1256e+02	0.1167e-01	
a	C(4) =	0.110652385632797e+00	0.2929e+00	0.1417e+02	0.2068e-01	
bb	C(5) =	-0.733666883810837e+00	0.3892e+00	0.2011e+02	0.1935e-01	
	C(6) =	-0.240712055489507e+01	0.1896e+00	0.1679e+01	0.1129e+00	
	C(7) =	0.154702780499544e+01	0.2200e+00	0.7964e+01	0.2763e-01	
	C(8) =	0.937279591540882e+00	0.6021e+00	0.2900e+02	0.2076e-01	
c	C(9) =	0.260859979315015e+00	0.4275e-02	0.3989e+01	0.1072e-02	
d	C(10) =	0.773174146647992e+00	0.6055e-01	0.1228e+02	0.4929e-02	
b	C(11) =	0.721993532572242e+00	0.7729e-01	0.9853e+00	0.7844e-01	
c	C(12) =	0.252719807732738e+00	0.3299e+00	0.6333e+01	0.5209e-01	
	C(13) =	-0.684847069135305e+00	0.2441e-01	0.6267e+01	0.3894e-02	
	C(14) =	-0.163037683043665e+01	0.2383e+00	0.7510e+01	0.3173e-01	
	C(15) =	-0.167572073277935e+01	0.1447e+00	0.1062e+02	0.1362e-01	

d	C(16)=	$-0.109838732122275e+01$	0.7076e+00	0.3256e+02	0.2173e-01
cc	C(17)=	$-0.782685857177615e+00$	0.3470e+00	0.3708e+01	0.9358e-01
e	C(18)=	$-0.274908030900022e+00$	0.4598e+00	0.5641e+01	0.8150e-01
	C(19)=	$-0.106323593166791e+01$	0.1634e+00	0.1633e+02	0.1001e-01
	C(20)=	$-0.811676687585208e-01$	0.1588e+01	0.1648e+02	0.9636e-01
	C(21)=	$0.353353103040516e+01$	0.5855e+00	0.1964e+01	0.2981e+00
dd	C(22)=	$-0.760921727843211e+00$	0.1885e+01	0.7290e+02	0.2585e-01
a	C(23)=	$-0.107652154992459e+00$	0.2401e+00	0.4315e+01	0.5564e-01
	C(24)=	$0.214073827536556e+01$	0.7777e+01	0.9677e+02	0.8037e-01
	C(25)=	$-0.461806267266914e+00$	0.7147e+01	0.1202e+03	0.5945e-01
b	C(26)=	$0.721895507040344e+00$	0.5002e+02	0.2761e+03	0.1812e+00

NB: $\left|c_{24}' - c_{24}''\right| = 4E\text{-}4$, $\left|c_{25}' - c_{25}''\right| = 3E\text{-}3$, $\left|c_{26}' - c_{26}''\right| = 3E\text{-}3$

(E)   Thus far: we have not supplied any hard information concerning actual CPU times. The following table gives a useful *overview* of this matter.

In interpreting these figures, bear in mind that each job consists of 2 parts: (a) the portion dealing with levels H1 & H2 ; (b) the portion stemming from any "blowups" that need to be made at level H3 .

A quick look at column 6 shows that the relative contribution of (a) & (b) can vary quite a bit from one job to another. [This fact needs to be taken into account when assigning time limits for the various jobs...]

	job category		α_1	machine	CPU time	total no. of H3 blowups	
1	N=4 R∈[5-ℓ, 10+ℓ] odd M = 10,11,12		.60	CRAY2(cft)	228 sec.	2+2+2=6	
2a	N=4 [20-ℓ, 25+ℓ] odd M = 17,18,19		.60	CRAY2(cft) XMP(cft77) YMP(cft77)	764 sec. 764 sec. 532 sec.	11+11+12=34	ON CRAY-2 time spent in part (a) $\approx$ 175 sec.
2b	N=4 [20-ℓ, 25+ℓ] even M = 18,19,20		.50	CRAY2(cft)	903 sec.	17+13+12=42	
3	N=5 [10-ℓ, 15+ℓ] even M = 23,24,25		.40	CRAY2(cft)	265 sec.	0+0+0 = 0	
4a	N=5 [20-ℓ, 25+ℓ] odd M = 28,29,30		.40	CRAY2(cft) XMP(cft77)	1494 sec. 1450 sec.	14+14+13=41	ON CRAY-2 time spent in part (a) $\approx$ 475 sec.
4b	N=5 [20-ℓ, 25+ℓ] even M = 28,29,30		.40	CRAY2(cft)	1097 sec.	11+6+8=25	
5	N=5 [45-ℓ, 50+ℓ] even M = 47,48,49		.40	CRAY2(cft)	2100 sec.	2+2+2 = 6	
6	N=5 [55-ℓ, 60+ℓ] even M = 56,57,58		.40	CRAY2(cft)	2008 sec.	0+0+0 = 0	
7	N=6 [20-ℓ, 25+ℓ] even M=31,32,33		.40	CRAY2(cft)	2714 sec.	19+26+24=69	

8	$N=7$ $[20-\ell, 25+\ell]$ odd $M=45,46,47$	.30	YMP(cft77)	2404 sec.	18+19+15=52
9	$N=7$ $[35-\ell, 40+\ell]$ even $M=62,63,64$	.30	YMP(cft77)	10594 sec.	41+41+33=115
10	$N=7$ $[45-\ell, 50+\ell]$ even $M=120,121,122$	.19	YMP(cft77)	5479 sec.	1+0+2=3
11	$N=5$ $49.875 \leq R \leq 51.00$ odd $M=48,50$ test run	.40	CRAY2(cft77) CRAY2(cft) XMP(cft77) YMP(cft77)	1270 sec. 1366 sec. 1370 sec. 966 sec.	7+7=14
12	$N=5$ $79.875 \leq R \leq 81.00$ odd $M=65,67$ test run	.45	CRAY2(cft77) CRAY2(cft) XMP(cft77) YMP(cft77)	2108 sec. 2274 sec. 2167 sec. 1529 sec.	8+10=18
	type $(\alpha_1, \alpha_2 \| \beta_1, \beta_2)$		H1=.025 , H2=.001 , H3=10^{-6}		$\ell \equiv \frac{1}{8}$

On the basis of this table, the following speed ratios are seen to apply:

$$\frac{\text{YMP(cft77)}}{\text{CRAY2(cft77)}} \cong 1.35 \ , \qquad \frac{\text{YMP(cft77)}}{\text{CRAY2(cft)}} \cong 1.45 \ , \qquad \frac{\text{YMP(cft77)}}{\text{XMP(cft77)}} \cong 1.42 \ . \quad [5]$$

In all:

$$\left\{ \begin{array}{c} 113 \\ 23 \\ 25 \end{array} \right\} \quad \text{of our type } (\alpha_1, \alpha_2 \| \beta_1, \beta_2) \text{ jobs used the } \left\{ \begin{array}{c} \text{CRAY2} \\ \text{XMP} \\ \text{YMP} \end{array} \right\} \qquad \bullet$$

The corresponding CPU totals were:

		odd				even			
	$N=4$	$N=5$	$N=6$	$N=7$	$N=4$	$N=5$	$N=6$	$N=7$	
CRAY2	3962	15222	4820	8319	9351	35337	22313	14326	113650 sec.
XMP	764	7720	8172	14931	0	0	0	6971	38558 sec.
YMP	532	2495	0	32571	0	0	0	49355	84953 sec.

---

[5] As usual, these figures represent a *composite* of both the memory and clock speeds.

(F) In our heuristic explanation of the residual effects, we noted that the same analysis could, in principle, be applied to more general Φ and that, there, the key indicator would be the size of γ^{-1}. Cf. (3.3) and the *analog* of (8.24), (8.21).

Earlier, in tables 13+16+18, we saw that unit-normalized pseudo cusp forms associated with $G_s(z;\rho)$ seem to be distinguished by the abnormally small size of their $\|\text{tail}\|$ -- at least when compared to any other [unit-normalized] Φ having singularities in $\mathcal{N}$, our usual tiny neighborhood of ρ.

In some sense: for these other Φ, one must somehow get a significantly bigger γ^{-1} which prevents $\underline{\Phi}(z)$ from ever being picked up (very sharply) in single-precision arithmetic.

The bracketed values in table 18 seem to suggest that there exists a kind of "graduation of levels" in γ^{-1} [controlling when any particular Φ gets picked up].

One would very much like to provide a rigorous basis for the "model" we've just described. At present, we are *unable* to do so.

It is possible, however, to make significant progress on at least one part of the problem; namely, the case where Φ has only one (logarithmic) singularity.

In short: we propose to examine $G_s(z;w_o)$ where $w_o \overset{\sim}{=} \rho$. In order for $G_s(z;w_o)$ to be even, we assume that $w_o \in \partial\mathcal{F}_N$.

Since γ varies continuously wrt w_o, it seems a bit *strange* that the unit-normalized Φ associated with such $G_s(z;w_o)$ aren't detected in tables 13+16+18.

Or are they????

We propose to show that $G_s(z;\rho)$ represents a kind of critical point wrt w_o. And: that the nearby $G_s(z;w_o)$ *are* (effectively) present.

To this end: let $s_o = \frac{1}{2} + iR_o$ $(R_o > 0)$ be a simple zero of $E(\rho;s)$. Assume that $\frac{1}{4} + R_o^2 \neq$ eigenvalue. Let $w(0) = \rho$, and $z = w(t)$ be a *nice* parametric representation for some subarc of $\partial\mathcal{F}_N$. Here $t \geq 0$.

Recall that $E(z;s)$ is holomorphic wrt s on a full neighborhood of s_o. Cf. [13, p.130(E)].

With this in mind, let $\mathcal{A}(t)$ be the corresponding zero of $E[w(t);s]$. For small t, this zero *must* be simple. (Use the argument principle!)

By the functional equation for $E(z;s)$, one immediately sees that $\text{Re}[\mathcal{A}(t)] = \frac{1}{2}$.

If $w(t)$ is real-analytic, so is $\mathcal{A}(t)$. In such cases, we automatically have:

$$(9.1) \qquad \frac{d\mathcal{A}}{dt} = -\frac{E_z[w(t);\mathcal{A}(t)]}{E_s[w(t);\mathcal{A}(t)]}\frac{dw}{dt} - \frac{E_{\bar{z}}[w(t);\mathcal{A}(t)]}{E_s[w(t);\mathcal{A}(t)]}\overline{\left(\frac{dw}{dt}\right)}$$

at least for small t.

When working with Fuchsian groups, one is (of course) free to replace the upper half-plane H by the unit disk U. We can therefore write $\mathfrak{C}_N \setminus U$ and suppose that ρ is (now) located at $z = 0$. On U, one has $\Delta f = (1-|z|^2)^2 f_{z\bar{z}}$. Cf. [13,p.14].

Let $\omega = \exp(2\pi i/N)$. Any *automorphic* function f on $\mathbb{C}_N \setminus U$ must satisfy $f(\omega z) \equiv f(z)$. Because of this invariance, the Taylor series development of $f(z)$ [wrt $z, \bar{z}$] can only include certain types of terms. To be more precise:

$$(9.2) \qquad f(z) = \begin{cases} f_{00} + f_{11} z\bar{z} + f_{22} z^2\bar{z}^2 + f_{30} z^3 + f_{03}\bar{z}^3 + O(|z|^5) & , \ N = 3 \\ f_{00} + f_{11} z\bar{z} + f_{22} z^2\bar{z}^2 + f_{40} z^4 + f_{04}\bar{z}^4 + O(|z|^5) & , \ N = 4 \\ f_{00} + f_{11} z\bar{z} + f_{22} z^2\bar{z}^2 + O(|z|^5) & , \ N \geq 5 \end{cases} \Bigg\} .$$

In particular: grad f must always *vanish* at $z = 0$. Note too that $f_{11} = (\Delta f)(0,0)$.

For functions like $E(z;s)$ or $F_n(z;s)$, one looks at Taylor series developments wrt $(z, \bar{z}, s - s_0)$ instead.

It is now easy to substitute (the) appropriate series expansions into (9.1). Bearing in mind that

$$f_{11} = 0 \qquad \text{for} \quad f = E(z; s_0) \qquad ,$$

we quickly establish that:

$$(9.3) \qquad \mathscr{A}(t) = \begin{cases} s_0 + O(t^3) & \text{for} \ \ N = 3 \\ s_0 + O(t^4) & \text{for} \ \ N \geq 4 \end{cases} \Bigg\} .$$

For $F_n[w(t); \mathscr{A}(t)]$, on the other hand, we have

$$(\Delta F_n)[0; s_0] = -\left(\tfrac{1}{4} + \mathcal{R}_0^2\right) F_n(0; s_0) \qquad ,$$

which is *not* generally zero. In this case, we find only that:

$$(9.4) \qquad F_n[w(t); \mathscr{A}(t)] = F_n(\rho; s_0)\left[1 - \left(\tfrac{1}{4} + \mathcal{R}_0^2\right)k^2 t^2\right] + O(t^3)$$

where $k = [2\mathrm{Im}(\rho)]^{-1}|w'(0)|$. Under quotients, however,

$$(9.5) \qquad \frac{F_n[w(t); \mathscr{A}(t)]}{F_1[w(t); \mathscr{A}(t)]} = \frac{F_n(\rho; s_0)}{F_1(\rho; s_0)}\left[1 + O(t^3)\right] \qquad .$$

It is in the sense of (9.3)–(9.5) that we say that $G_s(z; \rho)$ represents a (kind of) *critical point* wrt w_0.

We can always take t to be arc length (on H). Since $\mathcal{N}$ is small, so is the relevant t - range.

In view of (9.3) & (9.5), it is *not* too surprising that $G_s(z;w(t))$ doesn't show up explicitly. In some sense: $G_s[z;w(t)]$ stays "hidden" in the intrinsic fuzz. Cf. the error terms in tables 15 , 17 , 19.

The apparent sharpness *in* R is, of course, perfectly consistent with (9.3). In fact:

$$\left(\frac{\mathcal{L}}{2M}\right)^3 = 1.6E-5 \qquad for \qquad N = 3 , M = 20$$

$$\left(\frac{\mathcal{L}}{2M}\right)^4 = \begin{cases} 6.4E-7 & for \quad N = 4 , M = 25 \\ 1.8E-6 & for \quad N = 5 , M = 22 \\ 3.5E-6 & for \quad N = 6 , M = 20 \end{cases} \qquad .$$

[The preceding discussion is rather suggestive and *may* provide some hint as to how the more general cases should go. One also sees, in (9.3)+(9.5), the beginnings of a kind of variational theory wrt Φ ...]

— — — — — —

(G) For our next remark, let Γ_ε be a *real-analytic* family of Fuchsian groups with fixed signature. For simplicity, assume that the number of cusps is 1 . There is no loss of generality if we place the cusp at $i\infty$ and [then] normalize things in the usual way. Cf. [13,p.11].

In a situation like this: *one is inclined to say that*, within reason, *all the standard functions (like Eisenstein series and resolvent kernels) must vary real-analytically wrt* ε .

By adopting this hypothesis as one of our working tools, it becomes possible to derive an important limit formula which clarifies the destruction of cusp forms. [Cf. (9.9) and (9.13).]

In the analysis that follows, the reader is assumed to have a modest familiarity with [13,chapter 6] and [18] .

To get started, let Φ be a cusp form on $\Gamma_0 \setminus H$ with eigenvalue $\frac{1}{4} + R_0^2$ $(R_0 > 0)$ and multiplicity 1 . Assume that Φ is known to be destroyed by passing to $\Gamma_\varepsilon \setminus H$ (for small ε).

Define $\delta_\varepsilon(\lambda)$, $G(z;\lambda;\varepsilon)$, $A_\varepsilon(s)$, $\varphi_\varepsilon(s)$, $\psi_\varepsilon(s)$, $E_\varepsilon(z;s)$ in accordance with [13,pp.109-148, 231-232] . One knows that:

$$\delta_\varepsilon(\lambda) = A_\varepsilon(s)\,\psi_\varepsilon(s) \quad , \qquad \lambda = a(1-a) - s(1-s)$$

$$E_\varepsilon(z;s) = \frac{G(z;\lambda;\varepsilon)}{A_\varepsilon(s)} \quad , \qquad \varphi_\varepsilon(s) = \frac{A_\varepsilon(1-s)}{A_\varepsilon(s)}$$

$$\psi_\varepsilon(s) = \frac{1}{1-2a}\left[(1-s-a)B^{s-a} + \varphi_\varepsilon(s)(s-a)B^{1-s-a} \right]$$

$$\delta_\varepsilon(0) = A_\varepsilon(a) = \psi_\varepsilon(a) = 1 \qquad \bullet$$

By making a slight perturbation in B , we can (and do) assume that:

$$\psi_0(s_0) \neq 0 \qquad for \quad s_0 \equiv \frac{1}{2} + iR_0 \qquad \bullet$$

In view of [13,p.148(iii)] , the function $A_0(s)$ has a *simple zero* at s_0 . Choose δ so that $A_0(s) \neq 0$ for $0 < |s - s_0| < 5\delta$.

In everything that follows, we tacitly assume that $|s - s_0| \leq \delta$ and that δ , ε are kept sufficiently small.

We now apply the real analyticity of $A_\varepsilon(s)$ and $G[z;a(1-a)-s(1-s);\varepsilon]$. The

function $A_\varepsilon(s)$ will therefore have a simple zero s_ε very close to s_o.
Since $\overline{\Phi}$ is known to be destroyed, we must have $\mathrm{Re}(s_\varepsilon) < \frac{1}{2}$ for $\varepsilon \neq 0$.
Cf. [13,pp.148(iii),231].

Set:

$$s_\varepsilon \equiv \tfrac{1}{2} - \eta_\varepsilon + i R_\varepsilon \qquad and \qquad A_\varepsilon(s) \equiv (s - s_\varepsilon) V_\varepsilon(s)$$

wherein $V_\varepsilon(s) \neq 0$. The functions η_ε and R_ε are certainly real analytic.
Let:

$$(9.6) \qquad \eta_\varepsilon = b \varepsilon^{2p} + \sum_{j=1}^{\infty} b_j \varepsilon^{2p+j} \qquad , \qquad b > 0 \quad .$$

At the same time, expand:

$$G[z; a(1-a) - s(1-s); \varepsilon] = \sum_\ell B_\ell(z;\varepsilon)(s-s_\varepsilon)^\ell$$

$$= \sum_{k\ell} g_{k\ell}(z) \varepsilon^k (s-s_\varepsilon)^\ell \quad .$$

The functions B_ℓ are best expressed as complex integrals:

$$B_\ell(z;\varepsilon) = \frac{1}{2\pi i} \oint_{|s-s_o|=\delta} \frac{G[z; a(1-a)-s(1-s); \varepsilon]}{(s-s_\varepsilon)^{\ell+1}} ds \quad .$$

Since we're assuming *sensible* expansions wrt ε, the coefficients $g_{k\ell}$
are C^∞ functions on H with good Cauchy majorants [insofar as $|s-s_o| \leq \frac{1}{2}\delta$,
say].

Incidentally: note that

$$g_{oo}(z) = G[z; a(1-a) - s_o(1-s_o); 0] = A_o(s_o) E_o(z;s_o) = 0 \quad ,$$

by virtue of the fact that $E_o(z;s)$ has no poles along $\mathrm{Re}(s) = \frac{1}{2}$. Cf. [13, p.130(E)].

It follows that

$$(9.7) \qquad E_\varepsilon(z;s) = \frac{1}{V_\varepsilon(s)} \left[\sum_{k=1}^{\infty} g_{ko}(z) \frac{\varepsilon^k}{s-s_\varepsilon} + \sum_{\ell=1}^{\infty} B_\ell(z;\varepsilon)(s-s_\varepsilon)^{\ell-1} \right]$$

$$(9.8) \qquad \frac{s-s_\varepsilon}{\varepsilon} E_\varepsilon(z;s) = \frac{1}{V_\varepsilon(s)} \left[\sum_{k=1}^{\infty} g_{k0}(z) \varepsilon^{k-1} + \frac{s-s_\varepsilon}{\varepsilon} \sum_{\ell=1}^{\infty} B_\ell(z;\varepsilon)(s-s_\varepsilon)^{\ell-1} \right].$$

Suppose, for a moment, that $p = 1$. There are now 2 possibilities:

(a) $g_{10}(z) \not\equiv 0$;

(b) $g_{10}(z) \equiv 0$.

We begin with (a). Let $(s,\varepsilon) \longrightarrow (s_0,0)$ in such a way that

$$\text{Re}(s) \geq \frac{1}{2} \quad , \quad \frac{s-s_\varepsilon}{\varepsilon} \longrightarrow 0 \quad .$$

Since $\eta_\varepsilon \sim b\varepsilon^2$, this is certainly possible. Upon passing to the limit in (9.8), we see that:

$$\lim \frac{s-s_\varepsilon}{\varepsilon} E_\varepsilon(z;s) = \frac{1}{V_0(s_0)} \left[g_{10}(z) \right]$$

uniformly on H compacta.

Note that each $E_\varepsilon(z;s)$ is C^∞ wrt z. We can now apply the bounded convergence theorem as in $[13,\text{pp.}20,22(\text{line } 8)]$. This shows that:

$$\Delta g_{10} + \left(\tfrac{1}{4} + R_0^2\right) g_{10} = 0 \qquad \text{on} \quad H$$

and that g_{10} is Γ_0' - invariant .

But:

$$\frac{1}{V_0(s_0)} \int_0^1 g_{10}(x+iy) \, dx = \lim \frac{s-s_\varepsilon}{\varepsilon} \left[y^s + \varphi_\varepsilon(s) y^{1-s} \right] = 0$$

by virtue of the fact that $\varphi_\varepsilon(s) = O(1)$ for $\left\{ \text{Re}(s) \geq \frac{1}{2} \ , \ |s-s_0| < \delta \right\}$. Cf. $[13,\text{p.}156(\text{top})]$. A similar analysis shows that

$$\int_0^1 g_{10}(x+iy) e^{-2\pi iNx} dx = c_N \, y^{\frac{1}{2}} K_{iR_0}(2\pi|N|y)$$

for appropriate constants c_N. Accordingly ($[13,\text{p.}23(4.10)]$) :

$$g_{10}(z) = k_{10} \Phi(z) \qquad .$$

In summary then:

$$(9.9) \qquad \lim \frac{s-s_\varepsilon}{\varepsilon} E_\varepsilon(z;s) = k \bar{\Phi}(z) \qquad , \qquad k \neq 0$$

provided only that $(s, \varepsilon) \longrightarrow (s_0, 0)$ in such a way that

$$\operatorname{Re}(s) \geq \tfrac{1}{2} \qquad , \qquad \frac{s-s_\varepsilon}{\varepsilon} \longrightarrow 0 \qquad \bullet$$

⟦Taking $s = \tfrac{1}{2} + iR_\varepsilon$ is *one* illustration of this.⟧ $\quad (^6)$

It remains to consider (b). We propose to show that this case never arises.

To this end, let $g_{10}(z) \equiv 0$ and $\mathcal{S} = \tfrac{1}{2} + \eta_\varepsilon + iR_\varepsilon$. By applying (9.7), it is easily seen that:

$$(9.10) \qquad E_\varepsilon(z;\mathcal{S}) \Longrightarrow \frac{1}{V_0(s_0)} \left[\frac{g_{20}(z)}{2b} + B_1(z;0) \right] \qquad \text{on} \quad H \text{ compacta} \quad \bullet$$

Note too that:

$$(9.11) \qquad \begin{array}{ll} A_\varepsilon\left(\tfrac{1}{2} - \eta_\varepsilon + iR_\varepsilon\right) = 0 \quad , & A_\varepsilon\left(\tfrac{1}{2} + \eta_\varepsilon + iR_\varepsilon\right) \neq 0, \infty \\[2mm] \varphi_\varepsilon\left(\tfrac{1}{2} - \eta_\varepsilon + iR_\varepsilon\right) = \infty \quad , & \varphi_\varepsilon\left(\tfrac{1}{2} + \eta_\varepsilon + iR_\varepsilon\right) = 0 \end{array} \qquad \bullet$$

Cf. [13, pp. 148(iii), 130(B)(C), 166(12.8), 66(8.7)] . Here $\varepsilon \neq 0$.

Consider, now, any $s = \tfrac{1}{2} + h + it$ with $h > 0$, $|s - s_0| \leq \tfrac{1}{2}\delta$. By the Maass-Selberg identity:

$$\int_{\mathcal{F}_\varepsilon \cap \{y \leq Y\}} |E_\varepsilon(z;s)|^2 d\mu(z) \;+\; \int_{\mathcal{F}_\varepsilon \cap \{y \leq Y\}} |E_\varepsilon(z;s) - y^s - \varphi_\varepsilon(s) y^{1-s}|^2 d\mu(z)$$

$$= \frac{Y^{2h} - |\varphi_\varepsilon(s)|^2 Y^{-2h}}{2h} \;+\; 2\operatorname{Re}\left[\overline{\varphi_\varepsilon(s)} \frac{Y^{2it}}{2it} \right]$$

where $\mathcal{F}_\varepsilon$ is the obvious polygon for Γ_ε . Cf. [13, p. 200] . We have already noted that $\varphi_\varepsilon(s) = 0(1)$. Accordingly:

$(^6)$ For nonzero ε , the point s_ε represents a nontrivial pole of $E_\varepsilon(z;s)$.

$$\int\limits_{\mathcal{F}_\varepsilon \cap \{y \leq Y\}} |E_\varepsilon(z;s)|^2 \, d\mu(z) \;+\; \int_Y^\infty \int_0^1 \left| \sum_{n \neq 0} \varphi_{n\varepsilon}(s) y^{\frac{1}{2}} K_{s-\frac{1}{2}}(2\pi|n|y) e^{2\pi i n x} \right|^2 \frac{dx\,dy}{y^2}$$

$$= \; \frac{Y^{2h} - |\varphi_\varepsilon(s)|^2 Y^{-2h}}{2h} \;+\; O(1)$$

$$\int\limits_{\mathcal{F}_\varepsilon \cap \{y \leq Y\}} |E_\varepsilon(z;s)|^2 \, d\mu(z) \;+\; \int_Y^\infty \left[\sum_{n \neq 0} |\varphi_{n\varepsilon}(s) y^{\frac{1}{2}} K_{s-\frac{1}{2}}(2\pi|n|y)|^2 \right] \frac{dy}{y^2}$$

$$= \; \frac{Y^{2h} - |\varphi_\varepsilon(s)|^2 Y^{-2h}}{2h} \;+\; O(1)$$

$$\int\limits_{\mathcal{F}_\varepsilon \cap \{y \leq Y\}} |E_\varepsilon(z;s)|^2 \, d\mu(z) \;+\; \sum_{n \neq 0} |\varphi_{n\varepsilon}(s)|^2 \int_Y^\infty |K_{s-\frac{1}{2}}(2\pi|n|y)|^2 \frac{dy}{y}$$

$$= \; \frac{Y^{2h} - |\varphi_\varepsilon(s)|^2 Y^{-2h}}{2h} \;+\; O(1) \qquad \bullet$$

LEMMA. If Y is taken sufficiently large compared to R_0, it follows automatically that:

$$\int_Y^{2Y} |K_{\mathfrak{f}-\frac{1}{2}}(2\pi|n|y)|^2 \frac{dy}{y} \;\leq\; \int_Y^\infty |K_{\mathfrak{f}-\frac{1}{2}}(2\pi|n|y)|^2 \frac{dy}{y} \;\leq\; 2 \int_Y^{2Y} |K_{\mathfrak{f}-\frac{1}{2}}(2\pi|n|y)|^2 \frac{dy}{y}$$

for any $n \in \mathbf{Z} - \{0\}$ and $|\mathfrak{f} - s_0| < 1$.

Proof. Simply set $u = 2\pi|n|y$ and apply the standard asymptotic expansion of $K_\nu(u)$. Cf. [7, p.86(7)] or [13, p.22(iv)]. ∎

We now *fix* any Y consistent with the foregoing lemma. In short order:

$$\frac{Y^{2h} - |\varphi_E(s)|^2 Y^{-2h}}{2h} + O(1) = \int_{\mathcal{F}_E \cap \{y \le Y\}} |E_E|^2 d\mu + (1+\omega) \sum_{n \ne 0} |\varphi_{nE}|^2 \int_Y^{2Y} |K_{s-\frac{1}{2}}(2\pi |n| y)|^2 \frac{dy}{y}$$

$$\{\text{where } 0 \le \omega \le 1\}$$

$$= \int_{\mathcal{F}_E \cap \{y \le Y\}} |E_E|^2 d\mu + (1+\omega) \left[\int_Y^{2Y} \int_0^1 |E_E|^2 \frac{dx\,dy}{y^2} - \int_Y^{2Y} |y^s + \varphi_E(s) y^{1-s}|^2 \frac{dy}{y^2} \right]$$

$$= O(1) + \int_{\mathcal{F}_E \cap \{y \le Y\}} |E_E(z;s)|^2 d\mu(z) + (1+\omega) \int_Y^{2Y} \int_0^1 |E_E(z;s)|^2 \frac{dx\,dy}{y^2} .$$

Put $s = \mathcal{S}$ and apply (9.10)+(9.11). This leads to the relation:

$$\frac{Y^{2\eta_E}}{2\eta_E} = O(1)$$

which is clearly impossible. In other words: $g_{10}(z) \ne 0$, as promised.

An *alternate approach* consists of replacing (9.10) by a slightly more general assertion [wrt s] and *then* noting that:

(9.12) $$\frac{Y^{2h} - |\varphi_E(s)|^2 Y^{-2h}}{2h} = O(1) + \frac{1 - |\varphi_E(s)|^2}{2h} = O(1) + \frac{2\eta_E}{(h+\eta_E)^2 + (t-R_E)^2} \text{,}$$

by mimicking [13,p.163(top)] and using the fact that there are versions of [13,pp.157(middle), 162(line 6)] which are uniform in E . Compare [13,pp. 232(middle), 583(line 19), 571-574] .

The analysis for $p > 1$ is quite similar. In fact: let

$$L = \inf \{k : g_{k0}(z) \ne 0\} .$$

One easily verifies that $1 \le L \le 2p - 1$. Cf. the preceding *3* paragraphs.

⟦Note that $g_{L0}(z) \equiv$ (constant) $\bar{\Phi}(z)$, exactly like before. ⟧ By estimating the $|E_\varepsilon|^2$ integrals a bit more carefully, one quickly discovers that

$$\frac{1}{\eta_\varepsilon} \cong [\text{constant}] \left(\frac{\varepsilon^L}{\varepsilon^{ap}} \right)^2 \quad .$$

In other words: L = p.   With this relation in hand, the *final* result is clear; namely:

(9.13) $\displaystyle \lim \frac{s - s_\varepsilon}{\varepsilon^p} E_\varepsilon(z;s) = k \, \bar{\Phi}(z) \quad , \qquad k \neq 0 \; ,$

provided only that $(s, \varepsilon) \longrightarrow (s_0, 0)$ in such a way that

$$\mathcal{R}e(s) \geq \frac{1}{2} \quad , \qquad \frac{s - s_\varepsilon}{\varepsilon^p} \longrightarrow 0 \quad .$$

We are (of course) free to replace ε^p by the *branch* of $\sqrt{\eta_\varepsilon}$ which begins with $\sqrt{b}\,\varepsilon^p$.

The foregoing arguments are easily modified to take care of the ⟦formal⟧ case where $A_0(s) \neq 0$ on $\{\, |s - s_0| < 5\delta \,\}$.   Here there is no eigenvalue at $s_0(1 - s_0)$ and we simply find that:

$$\lim_{\varepsilon \to 0} E_\varepsilon(z;s) = E_0(z;s)$$

uniformly on compact subsets of $H \times \{\, |s - s_0| < \delta\}$. This assertion corresponds to [18, pp. 138(bottom)–139(top)]. ⟦The set $\mathcal{N}$ is defined on page *114* .⟧

See also: [43, section 7.1] .

(H) We finish up with a brief list of possibilities for further work.

(α) *Linear Algebra*.   Our subroutine for solving linear equations is nothing but standard Gauss elimination. It can *certainly* be improved.   For values of M bigger than 60 or so, it would probably be *best* to switch over to one of the optimized routines available in a standard library. ⟦This would help cut the CPU time! ⟧ Creative use of iterative techniques is another possibility.

(β) *Better Accuracy*.   It would certainly be useful to make at least a few runs with a *double-precision* version of the present code. This would provide us with a better understanding of both $G_s(z;\rho)$ and the various residual effects seen in §8. The difficulty in pursuing this idea stems *mainly* from the fact that most of the error terms in our KBES subroutines will need to be revised.

〖It virtually goes without saying that the issue raised in §9(F) paragraph 5 may prove to be *quite* important. For this reason: it makes sense to explore it using a variety of techniques...〗

(γ) *Additional Coefficients*. For arithmetic $\mathfrak{C}_N$, it would be *quite* useful to obtain many more Fourier coefficients ([7]) than we currently have. This can probably be done by implementing some version of H.Stark's method [41]. 〖Possible applications include Sato-Tate and the explicit computation of $L(s,\varphi)$.〗

(δ) *Other Groups*. It seems feasible to run similar R_n - experiments for more general Γ (and for more general multiplier systems). Cf. §1 footnote 3.

Can any *substitutes* be found for H.Stark's "bootstrap" method???

(ε) *Perturbation Expansions*. It would obviously be nice to have a *proof* of the real-analyticity hypothesis used in (G) !!

One's first reaction is that this problem *must* be "approachable" by means of some type of variational formalism.

In the case of E_ε , a slight variant of this idea does in fact work 〖thus *proving* (albeit in modified form) the real-analyticity needed to completely justify (9.9) , (9.13)〗. Cf. [46,§3.3]. ([8])

On a closely related matter, one *wonders*: is it possible to get a good enough 〖computational〗 hold on $E_\varepsilon(z;s)$ for generic $s \in \mathbb{C}$ to allow one to track s_ε by means of perturbation expansions and standard complex variable techniques?? (Such a capability would prove quite useful in a variety of contexts...)

REFERENCES

1. A.O.L.Atkin and J.Lehner, Hecke operators on $\Gamma_o(m)$, Math. Ann. 185(1970) 134-160.
2. N.L.Balazs and A.Voros, Chaos on the pseudosphere, Physics Rep. 143(1986) 109-240, especially chapter IX.
3. L.Bers, Uniformization, moduli, and Kleinian groups, Bull. London Math. Soc. 4(1972) 257-300.
4. J.M.Deshouillers, H.Iwaniec, R.S.Phillips, and P.Sarnak, Maass cusp forms, Proc. National Acad. Sci. USA 82(1985) 3533-3534.

([7]) for true φ 〈Cf. [52] for an update.〉
([8]) And: [30,pp.859(para 5) - 860(3.5) , 862(lines 9-13)] .   Note that a Neumann series can be used to compute $S(\Gamma)^{-1}$ on page 860. Compare [47].

5. P.Dienes, The Taylor Series, Dover Reprint, 1957.

6. H.M.Edwards, Riemann's Zeta Function, Academic Press, 1974, especially §§8.2,8.3 .

7. A.Erdélyi, et. al., Higher Transcendental Functions, vol. 2, McGraw-Hill, 1953,
   pages 86-88, especially pp.86(7), 87(18), 88(19)(20).

8. J.D.Fay, Fourier coefficients of the resolvent for a Fuchsian group, J. Reine
   Angew. Math. 294(1977) 143-203, especially pp.158-159.

9. G.H.Golub and C.F.Van Loan, Matrix Computations, Johns Hopkins Univ. Press, 1983.

10. L.Greenberg, Maximal groups and signatures, in Discontinuous Groups and Riemann
    Surfaces, Annals of Math. Studies No. 79, 1974, pp.207-226. See also: Bull.
    Amer. Math. Soc. 69(1963) 569-573.

11. H.Haas, Numerische Berechnung der Eigenwerte der Differentialgleichung $y^2\Delta u + \lambda u$
    $= 0$ für ein unendliches Gebiet im $\mathbb{R}^2$, Diplomarbeit, Universität Heidelberg, 1977,
    155pp.

12. E.Hecke, Mathematische Werke, Vandenhoeck & Ruprecht, Göttingen, 1959.

13. D.A.Hejhal, The Selberg Trace Formula for PSL(2,R), vol. 2, Springer Lecture Notes
    1001 (1983).

14. D.A.Hejhal, Eigenvalues of the Laplacian for PSL(2,Z) : some new results and
    computational techniques, in International Symposium in Memory of Hua Loo-Keng
    (ed. by Gong, Lu, Wang, Yang), Science Press and Springer-Verlag, 1991, vol. 1,
    pp.59-102.

15. D.A.Hejhal, Eigenvalues of the Laplacian for Hecke triangle groups, Technical
    Report 1990-16, Chalmers Univ. of Tech.(Sweden), 122pp.

16. D.A.Hejhal, Some observations concerning eigenvalues of the Laplacian and
    Dirichlet L-series, in Recent Progress in Analytic Number Theory, vol. 2,
    Academic Press, 1981, pp.95-110.

17. D.A.Hejhal, Zeros of Epstein zeta functions and supercomputers, in Proc. of the
    International Congr. of Math., Berkeley 1986, pp.1362-1384.

18. D.A.Hejhal, A continuity method for spectral theory on Fuchsian groups, in
    Modular Forms (ed. by R.A.Rankin), Ellis-Horwood, 1984, pp.107-140.

19. D.A.Hejhal and B.Berg, Some new results concerning eigenvalues of the non-Euclidean
    Laplacian for PSL(2,Z), Univ. of Minn. Math. Report No. 82-172, 1982, 7pp.

20. E.Hille, Analytic Function Theory, vol. 2, Ginn & Co., 1962.

21. A.Hurwitz, Über eine Reihe neuer Funktionen, welche die absoluten Invarianten
    gewisser Gruppen ganzzahliger linearer Transformationen bilden, Math. Ann. 20
    (1882) 125-134.

22. H.Iwaniec, Non-holomorphic modular forms and their applications, in Modular Forms
    (ed. by R.A.Rankin), Ellis-Horwood, 1984, pp.157-196, especially 178-180.

23. R.S.Lehman, Separation of zeros of the Riemann zeta function, Math. Comp. 20
    (1966) 523-541.

24. J.Lehner, <u>Discontinuous</u> <u>Groups</u> <u>and</u> <u>Automorphic</u> <u>Functions</u>, AMS Math. Surveys
    No. 8, 1964.

25. A.Leutbecher, Über die Heckeschen Gruppen $\mathfrak{G}(\lambda)$, Abh. Math. Sem. Hamburg 31
    (1967) 199-205.

26. T.Miyake, On automorphic forms on GL(2) and Hecke operators, Ann. of Math. 94
    (1971) 174-189.

27. H.Petersson, <u>Modulfunktionen</u> <u>und</u> <u>quadratische</u> <u>Formen</u>, Springer-Verlag, 1982.

28. H.Petersson, Über die eindeutige Bestimmung und die Erweiterungsfähigkeit von
    gewissen Grenzkreisgruppen, Abh. Math. Sem. Hamburg 12(1938) 180-199,
    especially 184(middle),193(Satz 4). Also: E.Schulenberg, same journal, 13(1939)
    144-199.

29. R.S.Phillips and P.Sarnak, On cusp forms for cofinite subgroups of PSL(2,R),
    Inventiones Math. 80(1985) 339-364.

30. R.S.Phillips and P.Sarnak, The Weyl theorem and the deformation of discrete
    groups, Comm. Pure Appl. Math. 38(1985) 853-866.

31. B.Randol, A remark on the multiplicity of the discrete spectrum of congruence
    groups, Proc. Amer. Math. Soc. 81(1981) 339-340.

32. R.A.Rankin, <u>Modular</u> <u>Forms</u> <u>and</u> <u>Functions</u>, Cambridge Univ. Press, 1977, pages
    245-342.

33. O.Rausenberger, Zur Theorie der Funktionen mit mehreren, nicht vertauschbaren
    Perioden, Math. Ann. 20(1882) 47-48.

34. P.Sarnak, On cusp forms, Contemp. Math. 53(1986) 393-407.

35. A.Selberg, Harmonic analysis and discontinuous groups in weakly symmetric
    Riemannian spaces with applications to Dirichlet series, J. Indian Math. Soc.
    20(1956) 47-87.

36. A.Selberg, Discontinuous groups and harmonic analysis, in <u>Proc.</u> <u>of</u> <u>the</u> <u>Internat.</u>
    <u>Congr.</u> <u>of</u> <u>Math.</u>, Stockholm 1962, pp.177-189.

37. G.Shimura, Differential operators and the singular values of Eisenstein series,
    Duke Math. J. 51(1984) 261-329.

38. G.Shimura, <u>Introduction</u> <u>to</u> <u>the</u> <u>Arithmetic</u> <u>Theory</u> <u>of</u> <u>Automorphic</u> <u>Functions</u>,
    Princeton Univ. Press, 1971, especially §§ 3.4 , 3.5 .

39. C.L.Siegel, <u>Advanced</u> <u>Analytic</u> <u>Number</u> <u>Theory</u>, 2nd edition, Tata Institute,
    Bombay, 1980.

40. D.Singerman, Finitely maximal Fuchsian groups, J. London Math. Soc. 6(1972)
    29-38.

41. H.Stark, Fourier coefficients of Maass waveforms, in <u>Modular</u> <u>Forms</u> (ed. by
    R.A.Rankin), Ellis-Horwood, 1984, pp.263-269.

42. K.Takeuchi, Arithmetic triangle groups, J. Math. Soc. Japan 29(1977) 91-106.
    Also: J. Fac.Sci. Univ. Tokyo 24(1977) 201-212.

43. A.B.Venkov, Spectral Theory of Automorphic Functions, Proc. Steklov Inst. Math. 153(1982) 163pp.

44. G.N.Watson, Asymptotic expansions of hypergeometric functions, Trans. Cambr. Philos. Soc. 22(1918) 277-308, especially p.289.

45. A.Winkler, Cusp forms and Hecke groups, J. Reine Angew. Math. 386(1988) 187-204.

46. S.Wolpert, The spectrum of a Riemann surface with a cusp, Univ. of Maryland preprint, 1990, 26pp.

47. T.Kato, Perturbation Theory for Linear Operators, Springer-Verlag, 1966, pages 400(thm 4.9), 408-409, 353, 423-426, 344(bot), 328(2.15).

48. A.Selberg, Collected Papers, vol. 1, Springer-Verlag, 1989.

49. G.W.Stewart, Introduction to Matrix Computations, Academic Press, 1973.

50. E.C.Titchmarsh, The Theory of the Riemann Zeta-Function, Oxford Univ. Press, 1951.

51. A.M.Turing, Some calculations of the Riemann zeta-function, Proc. London Math. Soc. 3(1953) 99-117, especially §3 and pp.111(middle)-112(middle).

52. D.Hejhal and S.Arno, On the Fourier coefficients of Maass waveforms for PSL(2,$\mathbf{Z}$), preprint, 1991.

53. M.Gutzwiller, Chaos in Classical and Quantum Mechanics, Springer-Verlag, 1990, chapter 15, especially figures 35,36,44,45.

54. M.Berry, Regular and irregular semiclassical wavefunctions, J.Physics A10(1977) 2083-2091.

ADDED IN PROOF (May 91).

(1) Our discussion has focussed mainly on 3 types of spectral objects; *viz.* true eigenfunctions (or cuspforms), pseudo cusp forms, and almost cusp forms [in the sense of (1.2)].

Strictly speaking, the concept of an almost cusp form does not arise unless the associated Teichmüller space is nontrivial. Cf. §9(G). *Be that as it may, however,* it should be remarked that none of our even production runs [with $N \stackrel{\leq}{=} 7$ and $R \stackrel{\leq}{=} 60$] ever encountered anything even remotely suggestive of (1.2), (9.9). This agrees with the philosophy expressed on pages 7(middle) and 28(paragraphs 3, 6).

[In order to detect anything like (1.2), one presumably needs to consider significantly larger R and/or N.]

(2) In connection with Tables 1-4, Scott Wolpert noticed the following very curious curious fact.

For *each* k , consider the multiplicative change in R_k as $N \longmapsto N + 1$. The resulting ratios display (what seems to be) a rather remarkable level of stability as k $\nearrow$. I.E.

N=3		N=4		N=5		N=6		N=7
9.5337	1.32	7.2209	1.12	6.4737	1.06	6.1206	1.03	5.9220
12.1730	1.28	9.5337	1.10	8.6368	1.05	8.1930	1.03	7.9339
14.3585	1.27	11.3177	1.12	10.1365	1.06	9.5337	1.04	9.1857
16.1381	1.33	12.1730	1.11	11.0156	1.05	10.5076	1.03	10.2292
16.6443	1.25	13.3102	1.10	12.0841	1.06	11.3659	1.04	10.9055
18.1809	1.27	14.3585	1.12	12.8513	1.06	12.1730	1.03	11.8036
19.4847	1.28	15.2740	1.09	14.0718	1.05	13.3786	1.04	12.8517
20.1067	1.25	16.1381	1.13	14.3079	1.06	13.5079	1.03	13.1539
21.4791	1.29	16.6443	1.10	15.1769	1.06	14.3585	1.04	13.8231
22.1947	1.27	17.4931	1.11	15.7599	1.07	14.7873	1.04	14.2401
23.2014	1.28	18.1809	1.12	16.2764	1.05	15.4832	1.03	15.0766
23.2637	1.26	18.4371	1.09	16.8910	1.05	16.1381	1.03	15.6258
24.4197	1.25	19.4847	1.10	17.7573	1.07	16.6443	1.05	15.8676
25.0509	1.25	20.1067	1.12	18.0314	1.06	16.9654	1.03	16.4936
26.0569	1.27	20.5302	1.10	18.6334	1.05	17.8207	1.04	17.1337
26.4470	1.26	21.0495	1.11	19.0117	1.06	18.0190	1.03	17.4714
27.2844	1.27	21.4791	1.11	19.3854	1.07	18.1809	1.03	17.6287
27.7759	1.25	22.1947	1.11	19.9622	1.05	19.0268	1.04	18.3243
28.5103	1.27	22.3749	1.09	20.5979	1.06	19.4847	1.04	18.7676
29.1376	1.26	23.2014	1.12	20.7456	1.06	19.5669	1.04	18.8689
29.5464	1.27	23.2637	1.09	21.2871	1.06	20.1067	1.03	19.4637
30.2790	1.26	24.0285	1.11	21.6756	1.06	20.4094	1.03	19.7701
30.4043	1.25	24.4197	1.10	22.1976	1.05	21.1087	1.04	20.3749
31.0565	1.24	25.0509	1.12	22.3994	1.04	21.4791	1.04	20.6554
31.9162	1.27	25.1193	1.09	23.0525	1.07	21.6127	1.03	20.9198
				23.4386	1.06	22.1003	1.04	21.3311
				23.5095	1.06	22.1947	1.03	21.4495
				24.0019	1.06	22.6711	1.03	21.9469
				24.2397	1.04	23.2014	1.05	22.1050
				24.6314	1.06	23.2637	1.02	22.7538
				25.0813	1.07	23.4602	1.03	22.7745
						24.2096	1.04	23.1868
						24.4197	1.04	23.5843
						24.9167	1.04	23.9027
						24.9526	1.04	24.0468
						25.0509	1.03	24.4186

To connect these figures with Weyl's asymptotic law [43, §6.7], bear in mind that $\lambda_k \equiv \frac{1}{4} + R_k^2$. The *expected* R_k - ratios should therefore be about

$$\sqrt{\frac{\mu(\mathcal{F}_{N+1})}{\mu(\mathcal{F}_N)}}$$

or 1.225, 1.095, 1.054, 1.035 (via page 9).

Needless to say: the agreement with experiment is quite good (*given* the relatively small size of our R_k *and* the presence of "boundary" terms in the relevant Weyl formula).

See also the related ideas in [18] (*and* [13,pp.528(bot), 599(85)]) *modulo* the symmetry $J(z) \equiv -\bar{z}$.

APPENDIX A

```
      PROGRAM xZKHU50
C     EXPERIMENTAL EIGENVALUE PROGRAM (D.A.HEJHAL//1989)
C     CRAY VERSION -- SINGLE PRECISION
C     for arbitrary Hecke groups: G(NGRP)
C     this program uses an adjustable Lagrange interpolation to
C     reduce the number of KBMAT calls;
C     it also uses TRIM indices (lifted from HSCALE.f);
C     and exploits ------
C     weighted Hecke indices
C     transversality controls
C     full KBMAT calls at H1
C     full SMAT calls at H2
C     full SMAT calls for MAG2 -- near "transverse" R values
C     to improve c(k) accuracy in NON-NOISY cases (cf. control #)
C     flexible program segments
C     control indices for noise/error/distortion
      IMPLICIT REAL(A-H,P-Z)
      PARAMETER(MAXDEG=51)
      PARAMETER(MG=16)
      PARAMETER(NMAX=104,NPAX=52,NCAX=52)
C     make certain that NMAX=2*NPAX=2*NCAX for later use!
      PARAMETER(NTRIM=2)
      PARAMETER(IPRINT=1)
      PARAMETER(LONG=55)
      PARAMETER(MAG=25)
      PARAMETER(LONGER=LONG*MAG)
      PARAMETER(LONGZER=LONG)
      PARAMETER(MAG2=1000)
      PARAMETER(MTW=9+2*MAG2)
      PARAMETER(NDEG=11)
C     make certain that NDEG is odd!!
      DIMENSION XA(NMAX),YA(NMAX),XP(NMAX),YP(NMAX)
      DIMENSION R3(0:LONG)
      DIMENSION R4(0:LONGER)
      DIMENSION RZ(LONGZER,NTRIM), RZT(LONGZER,NTRIM)
      DIMENSION ITRIM(NTRIM), LCOUNT(NTRIM)
      DIMENSION LKOUNT(NTRIM)
      DIMENSION GIANT(NMAX,NPAX,0:NDEG)
      DIMENSION U(NMAX,NPAX)
      DIMENSION UTMP1(NMAX,NPAX)
      DIMENSION UTMP2(NCAX,NPAX)
      DIMENSION CX1(NCAX)
      DIMENSION CDH(0:LONGER,NTRIM,NCAX)
      DIMENSION CDHX(0:LONGER,NTRIM,NCAX)
      DIMENSION CDAH(0:MTW,NCAX), WH(MTW), CDAHX(0:MTW,NCAX)
      DIMENSION R97(0:MTW)
      DIMENSION CDZ(LONGZER,NTRIM,NCAX)
      DIMENSION CDZX(LONGZER,NTRIM,NCAX)
      DIMENSION BU(NCAX)
      DIMENSION BUX(NCAX)
      DIMENSION CU(-2:3,NCAX)
      DIMENSION CUX(-2:3,NCAX)
      DIMENSION CUU(-2:3,2:4)
      DIMENSION VE(-1:1,NTRIM,NCAX), VEX(-1:1,NTRIM,NCAX)
      DIMENSION VDZ(LONGZER,NTRIM,NCAX)
      DIMENSION VDZX(LONGZER,NTRIM,NCAX)
      DIMENSION R2T(LONGZER,NTRIM), R3T(LONGZER,NTRIM)
      DIMENSION R4T(LONGZER,NTRIM)
      DIMENSION VDX(LONGZER,NTRIM,NCAX)
      DIMENSION VDY(LONGZER,NTRIM,NCAX)
      DIMENSION VDXX(LONGZER,NTRIM,NCAX)
      DIMENSION VDYX(LONGZER,NTRIM,NCAX)
      DIMENSION A1(16),W(16)
      DIMENSION AD(MG),WD(MG)
```

```
      INTEGER*8 FAC(20)
      DIMENSION RF(20),RG(20)
      COMMON /DENNIS/ AD,WD,PI,RTM,PIH
      COMMON /DH2/ RF,RG
      COMMON /DH3/ XA,YA,XP,YP,PIXX
      COMMON /DH4/ NN1R,NN1L,NN2,NN3,PI2
      COMMON /DAH/ GIANT
C
C
      UDEG=FLOAT(NDEG)
      NA97=INT((.5E0)*(UDEG-1.0E0)+(1.0E-4))
C
C
      FAC(1)=1
      DO 10  J=2,20
         FAC(J)=J*FAC(J-1)
10    CONTINUE
      DO 12  J1=1,20
         RF(J1)=(1.0E0)/FLOAT(FAC(J1))
         RG(J1)=(1.0E0)/FLOAT(J1)
12    CONTINUE
      PI=(4.0E0)*ATAN(1.0E0)
      PIH=(.5E0)*PI
      PI2=(2.0E0)*PI
      A1(1)= .98940 09349 91650 E0
      A1(2)= .94457 50230 73233 E0
      A1(3)= .86563 12023 87832 E0
      A1(4)= .75540 44083 55003 E0
      A1(5)= .61787 62444 02644 E0
      A1(6)= .45801 67776 57227 E0
      A1(7)= .28160 35507 79259 E0
      A1(8)= .09501 25098 376374 E0
      A1(9)=-A1(8)
      A1(10)=-A1(7)
      A1(11)=-A1(6)
      A1(12)=-A1(5)
      A1(13)=-A1(4)
      A1(14)=-A1(3)
      A1(15)=-A1(2)
      A1(16)=-A1(1)
      W(1)= .02715 24594 117541 E0
      W(2)= .06225 35239 386479 E0
      W(3)= .09515 85116 824928 E0
      W(4)= .12462 89712 55534 E0
      W(5)= .14959 59888 16577 E0
      W(6)= .16915 65193 95003 E0
      W(7)= .18260 34150 44924 E0
      W(8)= .18945 06104 55068 E0
      W(9)=W(8)
      W(10)=W(7)
      W(11)=W(6)
      W(12)=W(5)
      W(13)=W(4)
      W(14)=W(3)
      W(15)=W(2)
      W(16)=W(1)
      MM=MG/16
      RTM=(1.0E0)/FLOAT(2*MM)
      DO 23  J3=0,MM-1
      DO 24  K3=1,16
         N3=16*J3
         WD(N3+K3)=W(K3)
         AD(N3+K3)=FLOAT(1+(2*J3))-A1(K3)
24    CONTINUE
```

```
  23      CONTINUE
  C
  C       INPUT DATA:
  C       in addition to PARAMETERS at top
  C
          NGRP=5
          CTXX=COS(PI/FLOAT(NGRP))
          STXX=SIN(PI/FLOAT(NGRP))
          PIXX=PI/CTXX
          R1=49.75E0
          R2=51.125E0
          NN1R=4
          NN1L=16
          NN2=4
          NN3=4
          DATA ITRIM /48,50/
          NR=2
          NC=26
          DO 3 J=1,NR
          DO 4 I=1,NC
            XQW=(.05E0)+(.75E0)*FLOAT(I-1)/FLOAT(NC-1)
            XA(I+(J-1)*NC)=XQW
            YQW=(.40E0)+(.05E0)*FLOAT(J-1)/FLOAT(NR-1)
            YA(I+(J-1)*NC)=YQW
  4       CONTINUE
  3       CONTINUE
          NRX=2
          NCX=26
          DO 5 J=1,NRX
          DO 6 I=1,NCX
            XQW=(.05E0)+(.75E0)*FLOAT(I-1)/FLOAT(NCX-1)
            XA(NPAX+I+(J-1)*NCX)=XQW
            YQW=(.42E0)+(.05E0)*FLOAT(J-1)/FLOAT(NRX-1)
            YA(NPAX+I+(J-1)*NCX)=YQW
  6       CONTINUE
  5       CONTINUE
  C
  C       END-OF-DATA
  C
          H1=(R2-R1)/FLOAT(LONG)
          H2=H1/FLOAT(MAG)
          H3=H2/FLOAT(MAG2)
          DO 5102 JJ=0,LONG
            R3(JJ)=R1+H1*FLOAT(JJ)
  5102    CONTINUE
          DO 5103 K=1,NMAX
            S=(XA(K)**2)+(YA(K)**2)
            XP(K)=-XA(K)/S
            YP(K)=YA(K)/S
  5103    CONTINUE
          DO 5104 JJ=0,LONGER
            R4(JJ)=R1+H2*FLOAT(JJ)
  5104    CONTINUE
  C
  C
  C
          DO 5301 JX=0,NDEG
          R=R3(JX)
          CALL KBMAT(R,U)
          DO 5302 K0=1,NPAX
          DO 5303 K=1,NMAX
            GIANT(K,K0,JX)=U(K,K0)
  5303    CONTINUE
  5302    CONTINUE
```

```
 5301    CONTINUE
C
C        giant do-loop follows:
C
         DO 74 J=0,LONG-NDEG
C
C
         IF(J.NE.0) THEN
         RR97=R3(J+NDEG)
         CALL KBMAT(RR97,U)
CDIR$    NEXTSCALAR
         DO 5401 I=0,NDEG-1
         DO 5402 K0=1,NPAX
         DO 5403 K=1,NMAX
           S=GIANT(K,K0,I+1)
           GIANT(K,K0,I)=S
 5403    CONTINUE
 5402    CONTINUE
 5401    CONTINUE
         DO 5502 K0=1,NPAX
         DO 5503 K=1,NMAX
           S=U(K,K0)
           GIANT(K,K0,NDEG)=S
 5503    CONTINUE
 5502    CONTINUE
         ENDIF
C
C
         DO 6101 JB=0,MAG
         SX=FLOAT(NA97)+FLOAT(JB)/FLOAT(MAG)
         CALL FLAG(SX,UTMP1)
         M=JB+(J+NA97)*MAG
         DO 6102 I4=1,NTRIM
C        prepare an adjusted copy of UTMP1
C        note that the throw-away row can be modified!!!!!
         NMAT=ITRIM(I4)-1
         DO 6123 K=1,NMAT
           UTMP2(K,NMAT+1)=-UTMP1(K,1)
 6123    CONTINUE
         DO 6124 K0=1,NMAT
         DO 6125 K=1,NMAT
           UTMP2(K,K0)=UTMP1(K,K0+1)
 6125    CONTINUE
 6124    CONTINUE
         CALL SMAT(UTMP2,NMAT,CX1)
           CDH(M,I4,1)=1.0E0
         DO 6128 K=2,NMAT+1
           CDH(M,I4,K)=CX1(K-1)
 6128    CONTINUE
C
C
         DO 6223 K=1,NMAT
           UTMP2(K,NMAT+1)=-UTMP1(NPAX+K,1)
 6223    CONTINUE
         DO 6224 K0=1,NMAT
         DO 6225 K=1,NMAT
           UTMP2(K,K0)=UTMP1(NPAX+K,K0+1)
 6225    CONTINUE
 6224    CONTINUE
         CALL SMAT(UTMP2,NMAT,CX1)
           CDHX(M,I4,1)=1.0E0
         DO 6228 K=2,NMAT+1
           CDHX(M,I4,K)=CX1(K-1)
 6228    CONTINUE
```

```
C
C
 6102    CONTINUE
 6101    CONTINUE
 74      CONTINUE
C
C        we now examine the Fourier coefficients
C        there is a GREAT deal of arbitrariness here;
C        fine-tuning may be necessary
C
         NII=MAG*(NA97)
         NFF=MAG*(LONG-NA97)
         NI=NII+2
         NF=NFF-2
C
C        giant do-loop follows:
C
         DO 7101 I=1,NTRIM
         NMAT=ITRIM(I)-1
         LKT=0
         LKTT=0
         DO 7102 JC=NI,NF-1
         DO 7601 N=-2,3
         DO 7602 K=2,NMAT+1
           CU(N,K)=CDH(JC+N,I,K)
           CUX(N,K)=CDHX(JC+N,I,K)
 7602    CONTINUE
 7601    CONTINUE
         DO 7701 L=-2,3
         DO 7702 L2=2,4
           CUU(L,L2)=CU(L,L2)-CUX(L,L2)
 7702    CONTINUE
 7701    CONTINUE
C        begin the detailed coefficient analysis
         IF((CUU(0,2)*CUU(1,2)).LT.(0.0E0)) THEN
           MSP=2
         ELSE IF((CUU(0,3)*CUU(1,3)).LT.(0.0E0)) THEN
           MSP=3
         ELSE IF((CUU(0,4)*CUU(1,4)).LT.(0.0E0)) THEN
           MSP=4
         ELSE
           MSP=0
         ENDIF
C        the end-statement is near 7102 (far below)!!!
         IF(MSP.NE.0) THEN
         S=CUU(0,MSP)/(CUU(0,MSP)-CUU(1,MSP))
         SA=1.0E0-S
         RTEMP=SA*R4(JC)+S*R4(JC+1)
         DO 7603 K=2,NMAT+1
           BU(K)=SA*CU(0,K)+S*CU(1,K)
           BUX(K)=SA*CUX(0,K)+S*CUX(1,K)
 7603    CONTINUE
         AY1=CU(-1,2)/CUX(-1,2)
         A2=CU(0,2)/CUX(0,2)
         A3=CU(1,2)/CUX(1,2)
         A4=CU(2,2)/CUX(2,2)
         AH=BU(2)/BUX(2)
         B1=CU(-1,3)/CUX(-1,3)
         B2=CU(0,3)/CUX(0,3)
         B3=CU(1,3)/CUX(1,3)
         B4=CU(2,3)/CUX(2,3)
         BH=BU(3)/BUX(3)
         C1=CU(-1,4)/CUX(-1,4)
         C2=CU(0,4)/CUX(0,4)
```

```
      C3=CU(1,4)/CUX(1,4)
      C4=CU(2,4)/CUX(2,4)
      CH=BU(4)/BUX(4)
C
      Z=1.0E0
      AY1=ABS(AY1-Z)
      A2=ABS(A2-Z)
      A3=ABS(A3-Z)
      A4=ABS(A4-Z)
      AH=ABS(AH-Z)
      B1=ABS(B1-Z)
      B2=ABS(B2-Z)
      B3=ABS(B3-Z)
      B4=ABS(B4-Z)
      BH=ABS(BH-Z)
      C1=ABS(C1-Z)
      C2=ABS(C2-Z)
      C3=ABS(C3-Z)
      C4=ABS(C4-Z)
      CH=ABS(CH-Z)
C
      Q1=AMIN1(A2,AH,A3)
      Q2=AMIN1(B2,BH,B3)
      Q3=AMIN1(C2,CH,C3)
      Q91=AMIN1(AY1,A2,AH,A3,A4)
      Q92=AMIN1(B1,B2,BH,B3,B4)
      Q93=AMIN1(C1,C2,CH,C3,C4)
C
      IF(Q91.GT.(.01E0)) THEN
      GOTO 7102
      ENDIF
C
      IF(Q92.GT.(.01E0)) THEN
      GOTO 7102
      ENDIF
C
      IF(Q93.GT.(.01E0)) THEN
      GOTO 7102
      ENDIF
C
      LKT=LKT+1
      RZ(LKT,I)=RTEMP
C
C     use LOCAL transversality to trim the list
C
      FG=CUU(1,MSP)-CUU(0,MSP)
      CA1=(CUU(-1,MSP)-CUU(-2,MSP))/FG
      CA2=(CUU(0,MSP)-CUU(-1,MSP))/FG
      CA4=(CUU(2,MSP)-CUU(1,MSP))/FG
      CA5=(CUU(3,MSP)-CUU(2,MSP))/FG
      Z89=AMIN1(CA1,CA2,CA4,CA5)
      IF(Z89.LT.(0.0E0)) THEN
C     look at inner differences vs. outer differences
        QDE=Q1-Q91
        IF(QDE.GT.(0.0E0)) THEN
        GOTO 7102
        ENDIF
        IF(Q1.GT.(.007E0)) THEN
        GOTO 7102
        ENDIF
        QDE=Q2-Q92
        IF(QDE.GT.(0.0E0)) THEN
        GOTO 7102
        ENDIF
```

```
          IF(Q2.GT.(.007E0)) THEN
          GOTO 7102
          ENDIF
          QDE=Q3-Q93
          IF(QDE.GT.(0.0E0)) THEN
          GOTO 7102
          ENDIF
          IF(Q3.GT.(.007E0)) THEN
          GOTO 7102
          ENDIF
       ENDIF
C
       QMMM=AMAX1(Q1,Q2,Q3)
       IF(QMMM.GT.(.01E0)) THEN
       GOTO 7102
       ENDIF
C
C
       LKTT=LKTT+1
C
C
       V21=(CUU(0,2)-CUU(-1,2))/H2
       V22=(CUU(1,2)-CUU(0,2))/H2
       V23=(CUU(2,2)-CUU(1,2))/H2
       V31=(CUU(0,3)-CUU(-1,3))/H2
       V32=(CUU(1,3)-CUU(0,3))/H2
       V33=(CUU(2,3)-CUU(1,3))/H2
       V41=(CUU(0,4)-CUU(-1,4))/H2
       V42=(CUU(1,4)-CUU(0,4))/H2
       V43=(CUU(2,4)-CUU(1,4))/H2
       AT1=ABS(V22-V21)
       AT2=ABS(V22-V23)
       AT3=ABS(V22)
       AT4=ABS(V32-V31)
       AT5=ABS(V32-V33)
       AT6=ABS(V32)
       AT7=ABS(V42-V41)
       AT8=ABS(V42-V43)
       AT9=ABS(V42)
       R2T(LKTT,I)=AMAX1(AT1,AT2)/AT3
       R3T(LKTT,I)=AMAX1(AT4,AT5)/AT6
       R4T(LKTT,I)=AMAX1(AT7,AT8)/AT9
C
C
C      minimize the Hecke index next;
C      use full SMAT calls for CDAH
C
C
       U27=FLOAT(MAG2)+(1.0E-6)
       MRG=INT(U27/(5.0E0))+1
       H2H=(.5E0)*H2
C
C
       DO 7171 JX=0,NDEG
       R=RTEMP+FLOAT(2*JX-NDEG)*H2H
       CALL KBMAT(R,U)
       DO 7172 K0=1,NPAX
       DO 7173 K=1,NMAX
          GIANT(K,K0,JX)=U(K,K0)
 7173  CONTINUE
 7172  CONTINUE
 7171  CONTINUE
C
C
```

```
           DO 7621 JBB=0,2*MRG
             R97(JBB)=RTEMP+FLOAT(JBB-MRG)*H3
             WHN=FLOAT(NA97)+(.5E0)
             SX=WHN+FLOAT(JBB-MRG)/FLOAT(MAG2)
             CALL FLAG(SX,UTMP1)
C          prepare an adjusted copy of UTMP1
C          the throw-away row should be the same as for CDH!!!
           DO 7631 K=1,NMAT
             UTMP2(K,NMAT+1)=-UTMP1(K,1)
 7631      CONTINUE
           DO 7633 K0=1,NMAT
           DO 7635 K=1,NMAT
             UTMP2(K,K0)=UTMP1(K,K0+1)
 7635      CONTINUE
 7633      CONTINUE
           CALL SMAT(UTMP2,NMAT,CX1)
               CDAH(JBB,1)=1.0E0
           DO 7637 K=2,NMAT+1
               CDAH(JBB,K)=CX1(K-1)
 7637      CONTINUE
           DO 7731 K=1,NMAT
               UTMP2(K,NMAT+1)=-UTMP1(NPAX+K,1)
 7731      CONTINUE
           DO 7733 K0=1,NMAT
           DO 7735 K=1,NMAT
               UTMP2(K,K0)=UTMP1(NPAX+K,K0+1)
 7735      CONTINUE
 7733      CONTINUE
           CALL SMAT(UTMP2,NMAT,CX1)
               CDAHX(JBB,1)=1.0E0
           DO 7737 K=2,NMAT+1
               CDAHX(JBB,K)=CX1(K-1)
 7737      CONTINUE
           CC2=CDAH(JBB,2)/CDAHX(JBB,2)
           CC3=CDAH(JBB,3)/CDAHX(JBB,3)
           CC4=CDAH(JBB,4)/CDAHX(JBB,4)
           Z=1.0E0
           WH2=ABS(CC2-Z)
           WH3=ABS(CC3-Z)
           WH4=ABS(CC4-Z)
C          the functional WH is adjustable!!!
           WH(1+JBB)=WH2+WH3+WH4
 7621      CONTINUE
           M98=(2*MRG)+1
           JBBM=ISMIN(M98,WH,1)
           JBBM=JBBM-1
           RZT(LKTT,I)=R97(JBBM)
           DO 7640 K=1,NMAT+1
           CDZ(LKTT,I,K)=CDAH(JBBM,K)
           CDZX(LKTT,I,K)=CDAHX(JBBM,K)
 7640      CONTINUE
C
C          compute velocity fluctuations
C
           DO 7641 N=-1,1
           DO 7642 K=2,NMAT+1
             VE(N,I,K)=(CU(N+1,K)-CU(N,K))/H2
             VEX(N,I,K)=(CUX(N+1,K)-CUX(N,K))/H2
 7642      CONTINUE
 7641      CONTINUE
           DO 7651 K=2,NMAT+1
             AT1=ABS(VE(0,I,K)-VE(-1,I,K))
             AT2=ABS(VE(0,I,K)-VE(1,I,K))
             AT3=VE(0,I,K)
```

```
              VDZ(LKTT,I,K)=AMAX1(AT1,AT2)
              VDX(LKTT,I,K)=AT3
              VDY(LKTT,I,K)=AMAX1(AT1,AT2)/ABS(AT3)
              BT1=ABS(VEX(0,I,K)-VEX(-1,I,K))
              BT2=ABS(VEX(0,I,K)-VEX(1,I,K))
              BT3=VEX(0,I,K)
              VDZX(LKTT,I,K)=AMAX1(BT1,BT2)
              VDXX(LKTT,I,K)=BT3
              VDYX(LKTT,I,K)=AMAX1(BT1,BT2)/ABS(BT3)
 7651    CONTINUE
C
C
         ENDIF
C
C
 7102    CONTINUE
         LCOUNT(I)=LKT
         LKOUNT(I)=LKTT
 7101    CONTINUE
C
C        now do the final print-out
C
         PRINT*,'    '
         PRINT*,'BASIC PARAMETERS:'
         PRINT 90, R1,R2
         PRINT 91, H1
         PRINT 92, H2
         PRINT 921, H3,MRG
         PRINT 93, NDEG
         PRINT 94, R4(NI),R4(NF)
 90      FORMAT(1X,F15.10,2X,F15.10)
 91      FORMAT(1X,'H1=',F15.12)
 92      FORMAT(1X,'H2=',F15.12)
 921     FORMAT(1X,'H3=',F15.12,2X,'MRG=',I6)
 93      FORMAT(1X,'DEGREE=',I3)
 94      FORMAT(1X,'ACTIVE RANGE:',F15.10,2X,F15.10)
C
C
         PRINT*,'   '
         PRINT 9411, NR, NC
 9411    FORMAT(1X,'ROWS:',I3,3X,'COLS:',I3)
         PRINT 9412, YA(1), YA(NPAX)
 9412    FORMAT(1X,'Y RANGE:',F6.3,3X,F6.3)
C
C
         PRINT*,'   '
         PRINT 9411, NRX,NCX
         PRINT 9412, YA(NPAX+1),YA(NMAX)
C
C
         DO 8101 I=1,NTRIM
         NMAT=ITRIM(I)-1
         PRINT*,'    '
         PRINT 95, ITRIM(I)
 95      FORMAT('ITRIM=',I4)
         PRINT*,'MAXIMAL LIST (level H2):'
         PRINT 96, (RZ(L,I),L=1,LCOUNT(I))
 96      FORMAT(1X,'R=',F15.10)
C
C
         PRINT*,'TRANSVERSE LIST (level H3):'
         PRINT 961, (RZT(L,I),L=1,LKOUNT(I))
 961     FORMAT('TR=',F15.10)
C
```

```
C
            IF(IPRINT.NE.0) THEN
            DO 8201 L=1,LKOUNT(I)
            PRINT 962, RZT(L,I),R2T(L,I),R3T(L,I),R4T(L,I)
  962       FORMAT('FOR R=',F15.10,3X,E11.4,2X,E11.4,2X,E11.4)
            C2=CDZ(L,I,2)
            C3=CDZ(L,I,3)
            C4=CDZ(L,I,4)
            C5=CDZ(L,I,5)
            C6=CDZ(L,I,6)
            C7=CDZ(L,I,7)
            C8=CDZ(L,I,8)
            C9=CDZ(L,I,9)
            C10=CDZ(L,I,10)
            C11=CDZ(L,I,11)
            C12=CDZ(L,I,12)
            C13=CDZ(L,I,13)
            C14=CDZ(L,I,14)
            C15=CDZ(L,I,15)
            C16=CDZ(L,I,16)
            C2X=CDZX(L,I,2)
            C3X=CDZX(L,I,3)
            C4X=CDZX(L,I,4)
            C5X=CDZX(L,I,5)
            C6X=CDZX(L,I,6)
            C7X=CDZX(L,I,7)
            C8X=CDZX(L,I,8)
            C9X=CDZX(L,I,9)
            C10X=CDZX(L,I,10)
            C11X=CDZX(L,I,11)
            C12X=CDZX(L,I,12)
            C13X=CDZX(L,I,13)
            C14X=CDZX(L,I,14)
            C15X=CDZX(L,I,15)
            C16X=CDZX(L,I,16)
            HH2=ABS(C2-C2X)
            HH3=ABS(C3-C3X)
            HH4=ABS(C4-C4X)
            HH5=ABS(C5-C5X)
            HH6=ABS(C6-C6X)
            HH7=ABS(C7-C7X)
            HH8=ABS(C8-C8X)
            HH9=ABS(C9-C9X)
            HH10=ABS(C10-C10X)
            HH11=ABS(C11-C11X)
            HH12=ABS(C12-C12X)
            HH13=ABS(C13-C13X)
            HH14=ABS(C14-C14X)
            HH15=ABS(C15-C15X)
            HH16=ABS(C16-C16X)
            PRINT 98, HH2,HH3,HH4,HH5
   98       FORMAT(1X,'HDIFFS:',F14.8,2X,F14.8,2X,F14.8,2X,F14.8)
            PRINT 98, HH6,HH7,HH8,HH9
            PRINT 98, HH10,HH12,HH14,HH16
            PRINT 99,(K,CDZ(L,I,K),VDZ(L,I,K),VDX(L,I,K),VDY(L,I,K),
     &      K=2,NMAT+1)
   99       FORMAT(3X,'C(',I3,')=',2X,E22.15,4X,E11.4,2X,E11.4,2X,E11.4)
            PRINT 99,(K,CDZX(L,I,K),VDZX(L,I,K),VDXX(L,I,K),VDYX(L,I,K),
     &      K=2,NMAT+1)
 8201       CONTINUE
            ENDIF
 8101       CONTINUE
            END
C
```

```
C
C
      SUBROUTINE KBMAT(R,U)
C     CRAY VERSION -- SINGLE PRECISION
      IMPLICIT REAL(A-H,P-Z)
      PARAMETER(NMAX=104,NPAX=52,NCAX=52)
      DIMENSION XA(NMAX),YA(NMAX),XP(NMAX),YP(NMAX)
      DIMENSION U1(NMAX,NPAX)
      DIMENSION U2(NMAX,NPAX)
      DIMENSION U(NMAX,NPAX)
      COMMON /DH3/ XA,YA,XP,YP,PIXX
      COMMON /DH4/ NN1R,NN1L,NN2,NN3,PI2
      S2=(5.0E0)/R
      T1=(1.0E0)/(3.0E0)
      T2=(2.0E0)/(3.0E0)
      S=AMIN1(T1,S2**T2)
      DO 6101 K=1,NMAX
      X1=XA(K)
      Y1=YA(K)
      X2=XP(K)
      Y2=YP(K)
CDIR$ NEXTSCALAR
      DO 100 N=NPAX,1,-1
        YN=(PIXX)*FLOAT(N)*Y1
        C=(ABS(R-YN))/R
        IF(C.LE.S) THEN
          CALL KBES2(YN,R,RKBES)
        ELSE IF(YN.LT.R) THEN
          CALL KBES1(YN,R,RKBES)
        ELSE
          CALL KBES3(YN,R,RKBES)
        ENDIF
        U1(K,N)=SQRT(Y1)*RKBES
  100   CONTINUE
      DO 110 N=NPAX,1,-1
        YN=(PIXX)*FLOAT(N)*Y2
        C=(ABS(R-YN))/R
        IF(C.LE.S) THEN
          CALL KBES2(YN,R,RKBES)
        ELSE IF(YN.LT.R) THEN
          CALL KBES1(YN,R,RKBES)
        ELSE
          CALL KBES3(YN,R,RKBES)
        ENDIF
        U2(K,N)=SQRT(Y2)*RKBES
  110   CONTINUE
 6101 CONTINUE
      DO 6102 K=1,NMAX
        X1=XA(K)
        X2=XP(K)
        DO 200 N=1,NPAX
          ZN1=(PIXX)*FLOAT(N)*X1
          ZN2=(PIXX)*FLOAT(N)*X2
          U11=U1(K,N)
          U22=U2(K,N)
          U(K,N)=U11*SIN(ZN1)-U22*SIN(ZN2)
  200   CONTINUE
 6102 CONTINUE
      RETURN
      END
C
C
      SUBROUTINE KBES1(YN,R,RKBES)
      IMPLICIT REAL(A-H,P-Z)
```

```
      PARAMETER(MG=16)
      DIMENSION AD(MG),WD(MG)
      DIMENSION RF(20),RG(20)
      COMMON /DENNIS/ AD,WD,PI,RTM,PIH
      COMMON /DH2/ RF,RG
      COMMON /DH4/ NN1R,NN1L,NN2,NN3,PI2
      DIMENSION U(MG),V(MG)
      DIMENSION VPRM(MG),T(MG),E(MG)
      DIMENSION E1(MG),E2(MG)
      DIMENSION A9(MG),B9(MG)
      DIMENSION Z(MG)
      N1=NN1R
      N2=NN1L
C     ADDITIONAL PARAMETERS:
          ZCRIT1=0.25E0
          ZCRIT2=0.25E0
          ETA=0.00E0
C
C
C     for ETA:   at level exp(-51)
C     .00    R<98            .33    R>492
C     .02    R>98            .50    R>1430
C     .05    R>113           .66    R>5831
C     .10    R>144           .80    R>31157
C     .20    R>238
C
      X=YN
      T8=R/X
      C25=SQRT((R-X)*(R+X))
      DD=(C25)/R
      RDD=(1.0E0)/DD
      U1=ALOG((R+C25)/X)
      C24=110.00E0/C25
      C23=SQRT(C24)
      S=T8*(U1-DD)
      C6=COS(X*S)
      S6=SIN(X*S)
      XXR=0.0E0
      XXL=0.0E0
*
      IF (R.LT.(40.0E0)) THEN
         DO 9381 I=2,7
         UT=U1+FLOAT(I)
         S3=SINH(UT)
         C3=COSH(UT)
         C=T8*UT-S
         S1=C/S3
         C74=SQRT((S3-C)*(S3+C))
         C1=C74/S3
         VT=(2.0E0)*ATAN(S1/(1.0E0+C1))
         TT=-X*C3*C1 + R*(PIH-VT)
         IF(TT.LT.(-55.0E0)) THEN
           CC23=FLOAT(I)
           GOTO 9382
         ENDIF
9381     CONTINUE
9382     H1=AMIN1(C23,CC23)/FLOAT(N1)
         H2=(DD-ETA*(DD**3))/FLOAT(N2)
*
*
      ELSE
         H1=AMIN1(C23,2.0E0)/FLOAT(N1)
         H2=(DD-ETA*(DD**3))/FLOAT(N2)
      ENDIF
```

```
*
CDIR$   NEXTSCALAR
        DO 300   J1=1,N1
            XJ=FLOAT(-1+J1)
            AA=XJ*H1
        IF(AA.GE.ZCRIT1) THEN
            GOTO 33
        ELSE
            GOTO 34
        ENDIF
33      DO 400   KK=1,MG
            U(KK)=U1+AA+(RTM)*H1*AD(KK)
            S3=SINH(U(KK))
            C3=COSH(U(KK))
            C=T8*U(KK)-S
            S1=C/S3
            C74=SQRT((S3-C)*(S3+C))
            C1=C74/S3
            V(KK)=(2.0E0)*ATAN(S1/(1.0E0+C1))
            T(KK)=-X*C3*C1+R*(PIH-V(KK))
            E(KK)=EXP(AMAX1(T(KK),-70.0E0))
            VPRM(KK)=(T8*S3-C*C3)/(C74*S3)
            E1(KK)=E(KK)*WD(KK)
            E2(KK)=E1(KK)*VPRM(KK)
400     CONTINUE
        GOTO 35
34      DO 402   KK=1,MG
            Z(KK)=AA+(RTM)*H1*AD(KK)
            ZT=Z(KK)
            AQ0=(ZT**14)*(RF(16)*(DD+RG(17)*ZT)+
      &          RF(18)*(DD+RG(19)*ZT)*ZT*ZT)
            AQ1=(ZT**10)*(RF(12)*(DD+RG(13)*ZT)+
      &          RF(14)*(DD+RG(15)*ZT)*ZT*ZT)+AQ0
            AQ2=(ZT**6)*(RF(8)*(DD+RG(9)*ZT)+
      &          RF(10)*(DD+RG(11)*ZT)*ZT*ZT)+AQ1
            AQ3=(ZT**2)*(RF(4)*(DD+RG(5)*ZT)+
      &          RF(6)*(DD+RG(7)*ZT)*ZT*ZT)+AQ2
            AQ4=RF(2)*(DD+RG(3)*ZT)+AQ3
            A9(KK)=AQ4*(RDD)
            BQ0=(ZT**14)*(RF(15)*(DD+RG(16)*ZT)+
      &          RF(17)*(DD+RG(18)*ZT)*ZT*ZT)
            BQ1=(ZT**10)*(RF(11)*(DD+RG(12)*ZT)+
      &          RF(13)*(DD+RG(14)*ZT)*ZT*ZT)+BQ0
            BQ2=(ZT**6)*(RF(7)*(DD+RG(8)*ZT)+
      &          RF(9)*(DD+RG(10)*ZT)*ZT*ZT)+BQ1
            BQ3=(ZT**2)*(RF(3)*(DD+RG(4)*ZT)+
      &          RF(5)*(DD+RG(6)*ZT)*ZT*ZT)+BQ2
            BQ4=(DD+RG(2)*ZT)+BQ3
            B9(KK)=BQ4*(RDD)
            C3T=1.0E0+ZT*DD*B9(KK)
            SM=1.0E0+(RDD*ZT)
            SN1=SM+A9(KK)*ZT*ZT
            SN2=(2.0E0*SM)+A9(KK)*ZT*ZT
            CXX=SQRT(A9(KK))*SQRT(SN2)
            S1=SM/(SN1)
            C1=ZT*(CXX/SN1)
            V(KK)=(2.0E0)*ATAN(S1/(1.0E0+C1))
            VX=-B9(KK)-(B9(KK)-A9(KK))*RDD*ZT
            VPRM(KK)=VX/(CXX*SN1)
            T(KK)=R*(PIH-V(KK)-C3T*C1)
            E(KK)=EXP(AMAX1(T(KK),-70.0E0))
            E1(KK)=E(KK)*WD(KK)
            E2(KK)=E1(KK)*VPRM(KK)
402     CONTINUE
35      X1=SSUM(MG,E1,-1)
        X2=SSUM(MG,E2,-1)
```

```
        XXR=XXR+(RTM)*H1*(X1*C6-X2*S6)
        IF(T(MG).LT.-55.0E0) THEN
            GOTO 3
        ENDIF
300     CONTINUE
CDIR$   NEXTSCALAR
3       DO 600  J2=1,N2
            XJ=FLOAT(-1+J2)
            AA=-XJ*H2
        IF(AA.LE.(-ZCRIT2)) THEN
            GOTO 53
        ELSE
            GOTO 54
        ENDIF
53      DO 700  KK=1,MG
            U(KK)=U1+AA-(RTM)*H2*AD(KK)
            S3=SINH(U(KK))
            C3=COSH(U(KK))
            C=T8*U(KK)-S
            S1=C/S3
            C74=SQRT((S3-C)*(S3+C))
            C1=-C74/S3
            V(KK)=PI-(2.0E0)*ATAN(S1/(1.0E0-C1))
            T(KK)=-X*C3*C1+R*(PIH-V(KK))
            E(KK)=EXP(AMAX1(T(KK),-70.0E0))
            VPRM(KK)=-(T8*S3-C*C3)/(C74*S3)
            E1(KK)=E(KK)*WD(KK)
            E2(KK)=E1(KK)*VPRM(KK)
700     CONTINUE
        GOTO 55
54      DO 702  KK=1,MG
            Z(KK)=AA-(RTM)*H2*AD(KK)
            ZT=Z(KK)
            AQ0=(ZT**14)*(RF(16)*(DD+RG(17)*ZT)+
     &          RF(18)*(DD+RG(19)*ZT)*ZT*ZT)
            AQ1=(ZT**10)*(RF(12)*(DD+RG(13)*ZT)+
     &          RF(14)*(DD+RG(15)*ZT)*ZT*ZT)+AQ0
            AQ2=(ZT**6)*(RF(8)*(DD+RG(9)*ZT)+
     &          RF(10)*(DD+RG(11)*ZT)*ZT*ZT)+AQ1
            AQ3=(ZT**2)*(RF(4)*(DD+RG(5)*ZT)+
     &          RF(6)*(DD+RG(7)*ZT)*ZT*ZT)+AQ2
            AQ4=RF(2)*(DD+RG(3)*ZT)+AQ3
            A9(KK)=AQ4*(RDD)
            BQ0=(ZT**14)*(RF(15)*(DD+RG(16)*ZT)+
     &          RF(17)*(DD+RG(18)*ZT)*ZT*ZT)
            BQ1=(ZT**10)*(RF(11)*(DD+RG(12)*ZT)+
     &          RF(13)*(DD+RG(14)*ZT)*ZT*ZT)+BQ0
            BQ2=(ZT**6)*(RF(7)*(DD+RG(8)*ZT)+
     &          RF(9)*(DD+RG(10)*ZT)*ZT*ZT)+BQ1
            BQ3=(ZT**2)*(RF(3)*(DD+RG(4)*ZT)+
     &          RF(5)*(DD+RG(6)*ZT)*ZT*ZT)+BQ2
            BQ4=(DD+RG(2)*ZT)+BQ3
            B9(KK)=BQ4*(RDD)
            C3T=1.0E0+ZT*DD*B9(KK)
            SM=1.0E0+(RDD*ZT)
            SN1=SM+A9(KK)*ZT*ZT
            SN2=(2.0E0*SM)+A9(KK)*ZT*ZT
            CXX=SQRT(A9(KK))*SQRT(SN2)
            S1=SM/(SN1)
            C1=ZT*(CXX/SN1)
            V(KK)=PI-(2.0E0)*ATAN(S1/(1.0E0-C1))
            VX=-B9(KK)-(B9(KK)-A9(KK))*RDD*ZT
            VPRM(KK)=VX/(CXX*SN1)
            T(KK)=R*(PIH-V(KK)-C3T*C1)
            E(KK)=EXP(AMAX1(T(KK),-70.0E0))
            E1(KK)=E(KK)*WD(KK)
```

```
          E2(KK)=E1(KK)*VPRM(KK)
  702     CONTINUE
  55      X1=SSUM(MG,E1,-1)
          X2=SSUM(MG,E2,-1)
          XXL=XXL+(RTM)*H2*(X1*C6-X2*S6)
          IF(T(MG).LT.-55.0E0) THEN
             GOTO 4
          ENDIF
  600     CONTINUE
  4       RKBES=XXL+XXR
C
C
          XXH=0.0E0
          IF(R.LT.(75.0E0)) THEN
          U0=U1-DD
          NH=16*(1+INT(U0))
          HH=U0/FLOAT(NH)
          DO 9711 JH=1,NH
            XJ=FLOAT(-1+JH)
            AA=XJ*HH
            DO 9722 KK=1,MG
            U(KK)=AA+(RTM)*HH*AD(KK)
            C3=COSH(U(KK))
            C8=COS(R*U(KK))
            T(KK)=X*C3-(PIH)*R
            E(KK)=EXP(AMAX1(T(KK),-70.0E0))*C8
            E1(KK)=E(KK)*WD(KK)
  9722      CONTINUE
            X1=SSUM(MG,E1,-1)
            XXH=XXH+(RTM)*HH*X1
  9711    CONTINUE
          ENDIF
          RKBES=RKBES+XXH
          RETURN
          END
C
C
          SUBROUTINE KBES2(YN,R,RKBES)
          IMPLICIT REAL(A-H,P-Z)
          PARAMETER(MG=16)
          DIMENSION AD(MG),WD(MG)
          DIMENSION RF(20),RG(20)
          COMMON /DENNIS/ AD,WD,PI,RTM,PIH
          COMMON /DH2/ RF,RG
          COMMON /DH4/ NN1R,NN1L,NN2,NN3,PI2
          DIMENSION U(MG),V(MG)
          DIMENSION VPRM(MG),T(MG),E(MG),E4(MG)
          DIMENSION A9(MG),B9(MG)
          N1=NN2
C         ADDITIONAL PARAMETER:
              UCRIT=0.125E0
          X=YN
          H=R-X
          G1=(4.0E0)*SQRT(3.0E0)
          G2=-1.0E0/3.0E0
          U12=G1*(R**G2)
          IF (R.LT.(120.0E0)) THEN
            U11=ALOG((8.0E0)+(200.0E0/R))
          ELSE IF (R.LT.(333.0E0)) THEN
            U11=1.5E0
          ELSE
            U11=U12
          ENDIF
          H1=U11/FLOAT(N1)
          XX1=0.0E0
CDIR$  NEXTSCALAR
```

```
          DO 300   J1=1,N1
              XJ=FLOAT(-1+J1)
              AA=XJ*H1
          IF(AA.GE.UCRIT) THEN
              GOTO 33
          ELSE
              GOTO 34
          ENDIF
33        DO 400   KK=1,MG
              U(KK)=AA+(RTM)*H1*AD(KK)
              S3=SINH(U(KK))
              C3=COSH(U(KK))
              S33=S3/(U(KK))
              S1=U(KK)/S3
              C21=SQRT((S33+1.0E0)*(S33-1.0E0))
              C1=(C21)/S33
              C6=COS(H*U(KK))
              S6=SIN(H*U(KK))
              V(KK)=(2.0E0)*ATAN(S1/(1.0E0+C1))
              VPRM(KK)=(S33 - C3)/(S3*C21)
              T(KK)=-X*C3*C1 + R*(PIH-V(KK))
              E(KK)=EXP(AMAX1(T(KK),-70.0E0))
              E4(KK)=E(KK)*(C6-S6*VPRM(KK))*WD(KK)
400       CONTINUE
          GO TO 35
34        DO 402 KK=1,MG
              U(KK)=AA+(RTM)*H1*AD(KK)
              UT=U(KK)
              AQ0=(UT**14)*RF(16)
              AQ1=(UT**10)*(RF(12)+RF(14)*UT*UT)+AQ0
              AQ2=(UT**6)*(RF(8)+RF(10)*UT*UT) + AQ1
              AQ3=(UT**2)*(RF(4)+RF(6)*UT*UT) + AQ2
              A9(KK)=RF(2)+AQ3
              BQ0=(UT**14)*RF(17)
              BQ1=(UT**10)*(RF(13)+RF(15)*UT*UT)+BQ0
              BQ2=(UT**6)*(RF(9)+RF(11)*UT*UT) + BQ1
              BQ3=(UT**2)*(RF(5)+RF(7)*UT*UT) + BQ2
              B9(KK)=RF(3)+BQ3
              C3=1.0E0+A9(KK)*(UT*UT)
              S33=1.0E0+B9(KK)*(UT*UT)
              V11=B9(KK)*(2.0E0+B9(KK)*UT*UT)
              V12=SQRT(V11)
              VPRM(KK)=(B9(KK)-A9(KK))/(V12*S33)
              S1=(1.0E0)/(S33)
              C1=U(KK)*(V12/S33)
              C6=COS(H*U(KK))
              S6=SIN(H*U(KK))
              V(KK)=(2.0E0)*ATAN(S1/(1.0E0+C1))
              T(KK)=-X*C3*C1 + R*(PIH-V(KK))
              E(KK)=EXP(AMAX1(T(KK),-70.0E0))
              E4(KK)=(C6-S6*VPRM(KK))*E(KK)*WD(KK)
402       CONTINUE
35        X1=SSUM(MG,E4,-1)
          XX1=XX1+(RTM)*H1*X1
          IF (T(MG).LT.-55.0E0) THEN
              GOTO 5
          ENDIF
300       CONTINUE
5         RKBES=XX1
          RETURN
          END
C
C
```

```
      SUBROUTINE KBES3(YN,R,RKBES)
      IMPLICIT REAL(A-H,P-Z)
      PARAMETER(MG=16)
      DIMENSION AD(MG),WD(MG)
      COMMON /DENNIS/ AD,WD,PI,RTM,PIH
      COMMON /DH4/ NN1R,NN1L,NN2,NN3,PI2
      DIMENSION U(MG),V(MG)
      DIMENSION T(MG),E(MG),E4(MG)
      N1=NN3
      X=YN
      RPIH=R*PIH
      S=R/X
      XC=SQRT((X+R)*(X-R))
      C=(XC)/X
      AL=(2.0E0)*ATAN(S/(1.0E0+C))
      XCRAL=XC + R*AL
      IF(RPIH-XCRAL.LT.-125.0E0) THEN
          XX11=0.0E0
          GO TO 7
      END IF
      C23=110.00E0/(XC)
      U11=SQRT(C23)
      IF (R.LT.(40.0E0)) THEN
        U22=ALOG((8.0E0)+(125.0E0/X))
      ELSE
        U22=2.0E0
      ENDIF
      H1=AMIN1(U11,U22)
      H1=H1/FLOAT(N1)
      XX1=0.0E0
CDIR$ NEXTSCALAR
      DO 300 J1=1,N1
          XJ=FLOAT(-1+J1)
          AA=XJ*H1
      DO 400 KK=1,MG
          U(KK)=AA+(RTM)*H1*AD(KK)
          S3=SINH(U(KK))
          C3=COSH(U(KK))
          S33=S3/(U(KK))
          S1=S*(U(KK)/S3)
          C1=SQRT((S33+S)*(S33-S)) / S33
          V(KK)=2.0E0*ATAN(S1/(1.0E0+C1))
          T97=-X*C3*C1-R*V(KK)
          T(KK)=T97+XCRAL
          E(KK)=EXP(AMAX1(T(KK),-70.0E0))
          E4(KK)=E(KK)*WD(KK)
  400     CONTINUE
      X1=SSUM(MG,E4,-1)
      XX1=XX1+(RTM)*H1*X1
      IF(T(MG).LT.-55.0E0) THEN
          GO TO 5
      END IF
  300 CONTINUE
  5   XX11=XX1*EXP(RPIH-XCRAL)
  7   RKBES=XX11
      RETURN
      END
C
C
      SUBROUTINE FLAG(SX,SV)
      IMPLICIT REAL(A-H,P-Z)
      PARAMETER(NMAX=104,NPAX=52,NCAX=52)
      PARAMETER(NDEG=11)
      PARAMETER(MAXDEG=51)
      DIMENSION GIANT(NMAX,NPAX,0:NDEG)
      DIMENSION SV(NMAX,NPAX)
      DIMENSION P(0:MAXDEG),D(0:MAXDEG)
```

```
          COMMON /DAH/ GIANT
          DO 2300 J=0,NDEG
            P(J)=1.0E0
            D(J)=1.0E0
          DO 2310  K8=0,NDEG
            IF(K8.NE.J) THEN
            P(J)=P(J)*(SX-FLOAT(K8))
            D(J)=D(J)*FLOAT(J-K8)
            ENDIF
2310  CONTINUE
2300  CONTINUE
          DO 3300 K0=1,NPAX
          DO 3310 K=1,NMAX
            SQQ=0.0E0
          DO 3320 JB=0,NDEG
            SQQ=SQQ+GIANT(K,K0,JB)*(P(JB)/D(JB))
3320  CONTINUE
            SV(K,K0)=SQQ
3310  CONTINUE
3300  CONTINUE
          RETURN
          END
C
C
          SUBROUTINE SMAT(U,N,C)
          PARAMETER(NMAX=104,NPAX=52,NCAX=52)
          DIMENSION U(NCAX,NPAX+1), C(NPAX)
          DO 1000 M=1,N
            TEMP=0.0E0
            MAXI=0
            DO 1100 J=M,N
              IF(TEMP.LT.ABS(U(J,M))) THEN
              TEMP=ABS(U(J,M))
              MAXI=J
              ENDIF
1100      CONTINUE
*         swap rows if necessary
            IF(MAXI.NE.M) THEN
              DO 1200 J=1,N+1
              T=U(MAXI,J)
              U(MAXI,J)=U(M,J)
              U(M,J)=T
1200        CONTINUE
            ENDIF
            TEMP=U(M,M)
            U(M,M)=1.0E0
            DO 1300 J=M+1,N+1
              U(M,J)=U(M,J)/TEMP
1300      CONTINUE
            DO 1400 J=1,N
              IF(J.NE.M) THEN
                T=U(J,M)
                U(J,M)=0.0E0
                DO 1500 K=M+1,N+1
                U(J,K)=U(J,K)-(T*U(M,K))
1500          CONTINUE
              ENDIF
1400      CONTINUE
1000  CONTINUE
          DO 2001 M=1,N
            C(M)=U(M,N+1)
2001  CONTINUE
          RETURN
          END
```

II.   EIGENVALUES OF THE LAPLACIAN FOR PSL(2,$\mathbf{Z}$) : SOME NEW RESULTS AND COMPUTATIONAL TECHNIQUES (★)

In Memory of Hua Loo-Keng

§1. <u>Introduction</u>. One of the most interesting problems in the Selberg trace formalism from the standpoint of *computation* is the explicit determination of discrete eigenfunctions of the automorphic Laplacian. Cf. [6,7,17,20] for the necessary theoretical background.

By virtue of its distinguished role in modern number theory, it seems reasonable to place particular emphasis on the *modular group* PSL(2,$\mathbf{Z}$) and its congruence subgroups.

It is well-known that such groups [arithmetic/noncompact/finite volume] have *both* a discrete and continuous spectrum. Part of the difficulty (numerically) stems from the fact that the bulk of the discrete spectrum lies "buried" well-within the continuous spectrum.

In this paper: we'll attack *only* the "discrete side" of the spectrum.

The continuous side leads to Eisenstein series, whose computational aspects have been recently discussed elsewhere [10].

Note that there are now *two* aspects to the problem:

(A) computation of the correct eigenvalues $\lambda_n = \frac{1}{4} + R_n^2$;

(B) determination of the corresponding eigenfunctions $\varphi_n(z)$.

In the arithmetic case, part (B) is basically synonymous with computing the Fourier coefficients of φ_n . Cf. [7,pp.256(4.7),585,282(3.9)].

Various papers have appeared (since the early 1970's) dealing with some combination of (A) and (B) for the case $\Gamma = $ PSL(2,$\mathbf{Z}$) . Cf. [4], [8], [12], [7,appendix C], [19], [23] and the *additional* references cited in [7,appendix C]. Note that [23] also treats several other groups. In all these papers: R is kept less than 25 or so.

Recent work in quantum chaos (and the Riemann zeta function) has suggested that it might prove interesting to try to attack problems (A) & (B) for the case of considerably larger R . Cf. [1,2,3,15,18,22]. Once again: arithmetic groups seem deserving of special attention.

Because of (increasing) difficulties with the computation of the K-Bessel function $K_{iR}(X)$, the number 25 (or so!!) has remained as a kind of "barrier" to further progress.

Recently, however, Enrico Bombieri & the author have developed a *new* program for the computation of $\exp(\frac{\pi}{2}R)K_{iR}(X)$ which works quite well for R all the way out to 75000 or more. The program uses the identity

(★) First published in the book, <u>International Symposium in Memory of Hua Loo-Keng</u> (ed. by S.Gong, Q.Lu, Y.Wang, L.Yang), Science Press and Springer-Verlag, 1991. Used with permission.

(1.1) $$K_{iR}(X) = \frac{1}{2} \int_R e^{-X\cosh t}\, e^{iRt}\, dt$$

followed by a *deformation of contour* very similar to that used in stationary phase. [1]

   This program played an essential role in our recent investigation of the distribution of zeros of Epstein zeta functions (by way of Eisenstein series). Cf. [9,10].

   Though (in that work) R was typically kept larger than 3000, there is *no* difficulty making several changes in the original CRAY code (readjusting error terms, etc.) so that $\exp(\frac{\pi}{2}R)K_{iR}(X)$ can be efficaciously computed for all $1 \overset{\le}{=} R \overset{\le}{=} 75000$, $\frac{1}{5} \overset{\le}{=} X < \infty$, say.

   One naturally *wonders* to what extent this new code allows one to *dispense* with the "R-barrier" previously encountered [at least for $\Gamma = \mathrm{PSL}(2,\mathbf{Z})$] .

   A look at [8,§6] shows that the answer to this question depends not only on the computation of $K_{iR}(X)$ but *also* (just as importantly!!) on the linear algebra associated with solving

(1.2) $$\sum_{n=1}^{N} c_n I_n(z_j,R) = 0 \quad , \quad 1 \overset{\le}{=} j \overset{\le}{=} N$$

for various batches of test points $\{z_1,\ldots,z_N\}$. Compare §2 below.

   Our *primary aim* in this paper is to report on the outcome of several new (super-computer) experiments in this direction.

   As will be seen: the results we obtain go well beyond those in [7,appendix C] . In short: one is now able to sample λ_n all the way out to $\lambda = 250000$ or so.

   Hua Loo-Keng was quite interested in the author's original paper [8] and encouraged him to pursue further work in this area. It therefore seems especially fitting to be able to dedicate this paper to his memory.

   The computer time for this work was provided by 3 sources:
(i) the Minnesota Supercomputer Institute (CRAY-2) ;
(ii) the National Science Foundation (XMP-48 at the San Diego Supercomputer Center) ;
(iii) the Swedish NFR (CRAY-1 at Linköping) .

   Additional support was received from *NSF Grant DMS 86-07958* and the math department at Chalmers Tekniska Högskola (in Sweden). I am also grateful to Mr. Barry Rackner of the Minnesota Supercomputer Center for his assistance with the various operating systems, and in running many of the jobs [particularly those on the XMP].

---

[1] The choice of contour hinges on whether $R > X$, $R \approx X$, $R < X$. The deciding factor is the size of $|R-X|^{3/2}/\sqrt{R}$. The middle case corresponds to the so-called *transitional* region. Cf. [21,pp.235,225,202] for the proper perspective.

§2. The Basic Procedure.   The approach we use is a slight *modification* of the one outlined in [8].

We are interested in computing nonholomorphic cusp forms (i.e. Maass wave forms) for $PSL(2,\mathbf{Z}) \backslash H$. Such functions can be represented in the form

(2.1) $$\varphi(x + iy) = \sum_{n=1}^{\infty} c_n y^{\frac{1}{2}} K_{iR}(2\pi n y) \left\{ \begin{array}{c} cos(2\pi nx) \\ \hline sin(2\pi nx) \end{array} \right\}$$

subject to the condition that:

(a) $\varphi(-\frac{1}{z}) = \varphi(z)$ for $z \in H$;

(b) $\sum_{n=1}^{\infty} \dfrac{c_n}{n^s} = \prod_p \dfrac{1}{1 - c_p p^{-s} + p^{-2s}}$;

(c) $|c_n| \leq d(n) n^{1/4}$.

The coefficients c_n are real; $d(n)$ is the usual divisor function; p means prime. To avoid any confusion, we explicitly remark that

(2.2) $$\Delta \varphi + (\tfrac{1}{4} + R^2)\varphi = 0$$.

In other words: $\frac{1}{4} + R^2$ is the *real* eigenvalue, not R .   For later use, we also write $z^* \equiv 1/\bar{z}$.

Condition (a) reflects the fact that $PSL(2,\mathbf{Z})$ is generated by $E(z) \equiv -\frac{1}{z}$ & $S(z) \equiv z + 1$ and that φ needs to be automorphic. The condition $\varphi(z+1) = \varphi(z)$ is automatically fulfilled.

Condition (b) says that φ is an eigenfunction of the Hecke operators:

(2.3) $$T_p[f] \equiv \frac{1}{\sqrt{p}} \left[f(pz) + \sum_{j=0}^{p-1} f(\tfrac{z+j}{p}) \right]$$.

In fact: $T_p[\varphi] = c_p \varphi$.

The generalized Ramanujan–Petersson *conjecture* asserts that $|c_n| \leq d(n)$. Condition (c) is a well-known (partial) result in this direction. Cf. [13,14,11] for further information.

The function φ is said to be "even" or "odd" according to whether *cos* or *sin* appears in (2.1). (Similarly for the corresponding R .)

Let $\mathcal{F}$ be the usual fundamental polygon for $PSL(2,\mathbf{Z}) \backslash H$. For each $z \in int(\mathcal{F})$, *note* that condition (a) can be rewritten in the form:

$$(2.4) \qquad \sum_{n=1}^{\tilde{}} c_n I_n(z_j', R) \; = \; 0 \qquad ,$$

The function $I_n(z,R)$ is an explicit combination of K-Bessel functions and sines or cosines. In the "even" case, one can take:

$$I_n(z,R) \equiv \sqrt{y^*} \, K_{iR}(2\pi n y^*) \cos(2\pi n x^*) \; - \; \sqrt{y} \, K_{iR}(2\pi n y) \cos(2\pi n x) \quad ,$$

The "odd" case uses "+" and *sin*. Here $z^* \equiv x^* + iy^*$.

Since $K_{iR}(X)$ is asymptotic to $\sqrt{\frac{\pi}{2X}} \, e^{-X}$ for $X \gg R$, the terms $I_n(z,R)$ begin to decay *exponentially fast* once n exceeds $\frac{R}{2\pi y^*}$ or so. Cf. ([1]) *loc. cit.* and ([2]).

The obvious temptation is to now take N "very large" and to try to solve

$$(2.5) \qquad \sum_{n=1}^{N} c_n I_n(z_j', R) \; = \; 0$$

over a batch of N randomly chosen testpoints $\{z_1,\ldots,z_N\} \subseteq \text{int}(\mathcal{F})$. Cf. [8,p. 102(top)]. ([3])

To allow for larger values of R [cf. ([2]) and (1.1)] , the I_n should be *pre*multiplied by $\exp(\frac{\pi}{2} R)$ from the very start. The number N should be chosen so that

$$(2.6) \qquad |I_\ell(z_k, R)| \; \lesssim \; \left(\begin{array}{c} \text{something like} \\ 10^{-9} \end{array} \right) \cdot \max_{\substack{1 \leq j \leq N \\ 1 \leq n \leq N}} |I_n(z_j, R)|$$

for every $\ell > N$ and $k \in [1,N]$. 10^{-9} can be replaced by 10^{-12} or 10^{-15} if greater accuracy is desired.

If several *precautions* are observed, it is *highly probable* that this (simple-minded) procedure will actually yield correct "answers."

([2]) Two additional facts should be kept in mind here. First: that $\left|K_{iR}(X)\right| \lesssim K_{\frac{1}{2}}(X)$ $= \sqrt{\frac{\pi}{2X}} \, e^{-X}$. Second that (for fixed X) the function $K_{iR}(X)$ *oscillates* ("quasi-trigonometrically") in an envelope having width roughly equal to $O(e^{-\frac{\pi}{2}R})$. The *average distance* between successive zeros of $K_{iR}(X)$ is $\frac{\pi}{\omega}$ where $\omega \equiv \log(\frac{2R}{X})$. Cf. [16,p.315] .

For later use: note that the corresponding *average* for the list of even (or odd) R_n is $\frac{12}{R}$ if $\Gamma = PSL(2,\mathbf{Z})$. See [20,§6.7] and [7,p.511] .

([3]) The y_j are kept *bounded* [to ensure that y_j^* stays away from 0] .

Four items need to be checked. Specifically:

(i) it is *essential* that the *same* R be obtained even when the points z_j are varied;

(ii) likewise for the (associated) c_n ;

(iii) the (purported) c_n should satisfy the *multiplicative relations* implicit in condition (b) [e.g. $c_4 = c_2{}^2 - 1$, $c_6 = c_2 c_3$, $c_8 = c_2{}^3 - 2c_2$, $c_{12} = c_4 c_3$] ;

(iv) the c_n should *also* satisfy (c) and (*hopefully*) the Ramanujan-Petersson conjecture.

From the standpoint of "probability", precaution (iii) is probably the most convincing.

In practice: (iv) tends to take care of itself after (i)-(iii).

The foregoing precautions are "tempered" by the fact that (2.5) is a *truncation* and that all (real) numbers have *finite* precision on a computer. In particular:

$$(2.7) \qquad \sum_{n=2}^{N} c_n I_n(z_j, R) = - I_1(z_j, R) \qquad , \quad 1 \overset{\le}{=} j \overset{\le}{=} N - 1$$

typically leads to nonsensical c_n [i.e. noisy/excessively large values] once n *exceeds* a certain bound determined by R, N , and the z_j . This bound will be loosely referred to as the "c_n hump." It is quite visible in numerical experiments. [4]

This state-of-affairs plainly shows that (i)-(iv) have to be taken with a grain of salt. As precautions they should be carried out *only* so far as the basic numerical configuration allows.

In [8] and [12] , R was determined by solving the equation

$$(2.8) \qquad det \left[I_n(z_j, R) \right]_{\substack{1 \le j \le N \\ 1 \le n \le N}} = 0 \qquad .$$

This equation is *not* sufficient all by itself. Spurious R-values crop up unless precautions (i)-(iii) are observed.

An *alternative* procedure would be to replace (2.5) by (2.7) at the outset and then determine R by imposing the constraint that

$$(2.9) \qquad c_4 + 1 - c_2{}^2 = 0 \qquad (say) \qquad .$$

It is *again* essential to adhere to precautions (i)-(iii).

The procedure using (2.8) is prone to *excessive* machine noise as R $\nearrow$, so we go with (2.7)+(2.9) instead.

---

[4] It is reasonable to expect that the c_n "hump" will scale something like $\frac{R}{2\pi y}$, where y is *some* number between $\min(y_k)$ and $\max(y_k)$. Though this is *not* completely accurate, it is good enough for a first approximation [assuming [3] and (2.6)] .

§3. <u>Some Informal Remarks Concerning Implementation of the Basic Procedure</u>.
Though the basic procedure is quite transparent, its *implementation* is strewn with
obstacles. In §§3-4, we briefly discuss several of the more important ones.

To begin with: observe that N has to be kept larger than something like

$$\frac{R}{2\pi y_{min}}$$, where $y_{min} \equiv \min(y_k^*)$.

This is good and bad. As R increases, the amount of matrix algebra implicit in
solving (2.7) goes up. But, in line with $(^4)$, the c_n "hump" *also* goes up. This means
that we *should* be able to "read off" significantly more coefficients (as a "reward" for
our greater effort).

This rather enticing state-of-affairs was, in fact, one of the *primary* motivations
for trying to implement the whole scheme.

But the "down" side is also quite apparent. At the outset, there are *two* immediate
concerns.

$\boxed{A}$ Since the y_j are kept bounded, the basic matrix

$$\left[\; I_n(z_j, R) \; \right]_{\substack{1 \leq j \leq N-1 \\ 2 \leq n \leq N}} \qquad \text{(called } J_R \text{ for short)}$$

can, in principle, become more & more singular (or "ill-conditioned") as $R \to \infty$,
$N \to \infty$. Cf. [5]. This causes problems in (2.7).

A machine like the CRAY2 gains *enormously* in speed by virtue of vectorization
[a kind of automatic "parallel processing" of those parts of the code requiring *only*
simple (*but repetitive*) arithmetic operations]. The "catch" is that the variables must
all be single-precision. Passing to double-precision [i.e. 28 places instead of the
usual 14-15] typically requires not 2 (or 4) but over 100 times as long. For this
reason: there are obvious advantages to trying to remain with single-precision in (2.7)
for as long as possible. $(^5)$

Bear in mind (too) that $c_4 + 1 - c_2^2 = 0$ is solved in effect by "hitting" (2.7)
over & over again with Gauss elimination on a suitable R-grid, and then looking for
changes in sign [$c_4 + 1 - c_2^2$ being regarded as a function of R].

In proceeding along these lines: it is essential to try to arrange the z_j in
some "reasonably favorable" configuration so that the *conditioning* of J_R is optimized
-- and so that *single* precision does indeed suffice.

Though precaution (iii) will "let us know" when we've arrived, there are no *apriori*
guarantees. One can only HOPE for the best.

We'll return to this point later.

$(^5)$ Unless, of course, time and money are no object.

[B] The *second* difficulty is more mundane: namely the repetitive computation of *all* those Bessel functions in J_R .

To solve $c_4 + 1 - c_2^2 = 0$, the basic R-grid has to be kept "finer" than a *small fraction* of the average distance between successive R_n . (Otherwise pairs of "nearby" R_n will be missed.) This means that the basic ΔR [in (2.7)] has to be kept at something like:

(3.1) $\dfrac{1}{50}\left(\dfrac{12}{R}\right)$ or $\dfrac{1}{100}\left(\dfrac{12}{R}\right)$ (or less!)   •

Cf. (^{2}). This grid can (*then*) be further refined on those intervals (actually) containing a change-of-sign.

This process clearly involves a lot of computation. One naturally thinks of trying to use Lagrange interpolation (or something similar!) in an attempt to save time. [The functions $K_{iR}(X)$ are, after all, *holomorphic* with respect to R.]

This idea is actually quite fortuitous. To see this, let $X_{min} = 2\pi y_{min}$. For holomorphic functions, one is inclined to say (loosely speaking) that Lagrange interpolation is *reasonable* so long as the degree is large and the grid size is less than a small fraction of the *average distance* between the sucessive "peaks and valleys" on the graph. For the Bessel functions appearing in J_R , this distance is *not less* than $\dfrac{\pi}{\log\left(\frac{2R}{X_{min}}\right)}$. Cf. (^{2}). Since

(3.2) $\dfrac{1}{10}\left[\dfrac{\pi}{\log\left(\frac{2R}{X_{min}}\right)}\right] \gg \dfrac{1}{50}\left(\dfrac{12}{R}\right)$ typically ,

we *should* be able to reap a *very* significant savings (at least if life is reasonable). Whether life is "reasonable" or not can (of course) be *checked* by making direct "spot-checks" in various R - neighborhoods. In our case: it was determined that [for degree 11] and a coarse grid like the LHS of (3.2), we were achieving $10 \sim 12$ significant figures all the way out to R = 1000 .

— — — — — —

Concerns [A] and [B] suggest that there should exist *three* basic length scales

H1 > H2 > H3

in the actual program. (^{6})

—————————

(^{6}) at least for starters!

H1   is the "coarse" grid on which the initial computation of all the Bessel functions takes place.

H2   is the "finer" scale on which -- by Lagrange interpolation -- the matrices J_R
are all assembled and then "manipulated" [ala (2.7)] to locate changes-of-sign for
$c_4 + 1 - c_2^2$.

H3   is the "final" scale reflecting the accuracy we *ultimately* hope to achieve for
R   when solving (2.9).

To describe the passage from H2 to H3 , it is necessary to supply a bit more
detail concerning the actual *mechanics* of the code.

Prior to doing this, we should perhaps stress that, in designing the code, our basic
attitude was [and still is] that we would think of the program as being *only* experimental.

A number of obstacles had to be overcome in an *ad hoc* way. The "evolving" program
gradually needed to take on various parameters and adjustable components. We have *not*
yet made any systematic attempt to optimize *all* aspects simultaneously. [It takes
nontrivial cpu time to run the necessary tests.]

On the other hand: there are *only* a finite number of eigenvalues from R = 1 out to
R = 1000 (say). Once these are known to 6 decimal places (or whatever), further "stream-
lining of the code" could well be moot.

We have therefore sought to maintain a proper *balance* , philosophically, in our use
of:

(i) educated guessing ;

(ii) small-scale empirical tests ;

(iii) rigorous theoretical estimates (and error analyses) ;

(iv) sheer luck.

§4. <u>Further Remarks</u>. When designing a code that is expected to require *hours*
(as opposed to minutes) of CRAY time, it is wise to make preliminary tests ([7]) in at
least *one* TYPICAL regime. Features that are totally insignificant (or "invisible")
at R = 25 may , for instance, grow large enough to virtually "destroy" a code at
R = 1000 .

In the present case, we decided to fix a modest value of R (125)   and then see
if we could *at least* get the code to work reasonably well in a small neighborhood of
this value.

The number 125 was a compromise: large enough to be "interesting" yet small
enough to allow the testing to be done fairly quickly.

Our guess was that, with a bit of "fine-tuning", the basic procedure would actually
work -- in *single* precision. [Verifying this *hunch* (about s.p.) was the *other* primary

----

([7]) or sample runs

motivation for pursuing these experiments. The first was the *hope* of obtaining more c_n as R increased.]

The initial tests at 125 were very discouraging until (in desperation) we tried a few "last" z_j - configurations and a slightly different R-interval. We *finally* found two R-values where precaution (iii) worked beautifully.

The ensuing optimism was tempered by the *disturbing fact* that concern $\boxed{A}$ was apparently very real. Depending on the z_j - batch used, the matrix J_R could either be "well-conditioned" or a "disaster." It is (to be sure) eminently reasonable to *expect* that $\det(J_R)$ will have zeros. Just like (2.8). But one would *hope* that, for generic z_j , the two zero-sets would be "reasonably disjoint."

In cases (*or regions*) where J_R is a "disaster", one will typically miss many "true" R because of all the "static."

The idea of automatically testing *several* z_j - batches to "increase the complete-ness probability" now becomes rather clear.

Something akin to this actually seems quite essential — *at least if one still hopes to use* only *single-precision.*

On the matter of z_j - configurations, our naive *guess* was that the z_j needed to be "spread out" as far as possible -- consistent with (^{3}). The interesting (and ini-tially *deceptive*) feature was that this idea did *not* work -- at least for rectangular configurations.

In practice: the "best" J_R seemed to arise from cases having $\max(y_k)$ fairly small [e.g. 2] with a *minimal* number of vertical columns [again like 2].

There were *also* some difficulties with (2.9) itself which needed to be addressed in the code.

It was often observed that the zeros of $c_4 + 1 - c_2^2$ were "fuzzy." That is: even when H2 was consistent with (3.1), there frequently appeared "strings" of 2 or more successive intervals *each* containing (*according* to the machine) a change-of-sign. The explanation for this (^{8}) is obviously one of machine noise. Some z_j - batches were [in line with §3A] *much worse* than others in this regard.

This type of "fuzziness" was already noted by P.Cartier in connection with (2.8) -- but (2.8) is generally quite a bit "noisier" than (2.9).

One naturally attempts to get a "better grip" on any R (using *only* the H2-level data) by applying a linear interpolation or something similar. Each interval containing a change-of-sign thus leads to a number called R_{temp} .

In the case of fuzzy zeros, one obtains *strings* of R_{temp}'s .

To solve the problem of eliminating the "phonies", we kept in mind the "dictum" (or basic necessity) of *always* testing several z_j batches [to maximize the

---

(^{8}) when R is still comparatively small

probability of completeness]. There was room here to be a bit wasteful, so we simply declared as "unsafe" any case where the crossing associated with R_{temp} was *not* locally transverse [on a scale of several H2 intervals] -- and then proceeded *no* further with this value (beyond printing it). This seemed to work reasonably well.

To pass from level H2 to H3, one effectively needs to "blow up" small neighborhoods of R_{temp} .

Any type of "refinement process" is bound to take time [especially if $\frac{H2}{H3}$ is large!] , so it obviously pays to eliminate *beforehand* any "case" strongly suspected of being spurious. By that we mean any R_{temp} where one of the Hecke relations beyond (2.9) [e.g. $c_2 c_3 = c_6$] is *already* violated ([9]) at the level of H2 .

For $R \approx 125$, this "weeding process" was already a virtual necessity in every z_j- batch.

At this point: another significant problem occurs. Prompted by certain irregularities in deciding whether the "weeding test" could actually be trusted, we performed a variety of tests on the coefficients c_n . It was found (*especially* for larger R!!) that the c_n tend to "act" like *continuous* but *nowhere* differentiable functions of R as far as the *machine* goes. This is due to machine noise in solving (2.7). As n gets closer to the c_n hump , the coefficient "velocities" (wrt R) tend to grow ever larger. ([10]) The corresponding c_n are thus *very* sensitive.

This observation is actually relevant even at $R \approx 13.779751$ (the *first* even R). There the c_n "hump" was located at about 7 . This makes testing (iii) a bit tricky. [One has to be virtually on top of the true R to see the c_6 or c_8 relations begin to work. But, since (in this case), the velocity *fluctuations* remain fairly small, one *does* eventually get there. And *all* in single precision. Cf. §10 below.]

The " c^0-c^1 problem " was addressed by introducing a "filter" instead of employing any kind of bisection or *regula falsi* to solve (2.9).

The basic idea is to *re*compute all the Bessel functions on an appropriate H2 grid around R_{temp} , and then use Lagrange interpolation to prepare the matrices J_R along a suitable H3 grid containing R_{temp} . The "true" R is then obtained by *repeatedly* solving (2.7) and seeking to *minimize* a test functional like:

$$(4.1) \qquad \left| c_4 + 1 - c_2^2 \right| \quad + \quad \left| c_6 - c_2 c_3 \right| \quad + \quad \left| c_8 - c_2^3 + 2c_2 \right| \qquad .$$

By keeping H3 sufficiently small ([11]), this process seems to act as a reasonably good *filter* for whatever c_n noise [or "static"] may be present. Because regula falsi fails,

---

([9]) sufficiently badly
([10]) "Velocities" like 10^5 or 10^6 are quite common (even at the level of H2).
([11]) this condition is essential

one is "taking significant chances" if the filter is applied on an interval
$\{|R - R_{temp}| \leqq \gamma(H2)\}$ with γ *significantly less* than $\frac{1}{2}$ or so.

In practice, it didn't seem to make much difference which functional was used
in (4.1) so long as it was reasonable. We generally used

$$(4.2) \qquad \left| \frac{c_2^2 - 1}{c_4} - 1 \right| \; + \; \left| \frac{c_2 c_3}{c_6} - 1 \right| \qquad .$$

One way to speed the process up -- as regards the *repeated* solving of (2.7) --
is to use the same type of code but to *assume* that the c_n can be legitimately computed
at level H3 by mere *linear* interpolation of the coefficients *already* obtained at
level H2. Some surprisingly good answers can be obtained in this way. Because of the
$c^0 - c^1$ problem, this method is certainly *not* rigorous -- but does have the advantage
of locating *some* of the true R-values (to a few less decimal places) rather quickly.([12])
The more refined version of the code can then be used if greater accuracy is desired.

One can also use this "shortcut" to help determine which cases are most "stable" --
or to search for missing R-values.

This completes our remarks on the implementation of §2 except for noting one
further "curiosity."

Different machines -- *and even different compilers* -- can ([13]) give rise to different
noise [or "garbage bits"] when the matrices J_R are prepared. For this reason: cases
which were "triggered" at level H2 in *one* setting can be completely *missed* in another.
This behavior reflects 2 things; namely
(i) the intrinsic variability in the c_k noise level ; and
(ii) the fact that the noise is filtered only AFTER the "trigger" is "activated" at
level H2 by a suitably transverse crossing of $c_4 + 1 - c_2^2$.

We saw *many* examples of this irritating *anomaly* as R went beyond 125 . ([14])

With the code structured the way it is now, the notion of trying to increase the
"completeness probability" is obviously *very* fundamental. At present: one simply
does not have any kind of *theoretical* guarantees [or, for that matter, any *apriori*
knowledge of optimal z_j]. ([15]) One can only be guided by knowledge of the average

---

([12]) at least for moderate R !! Beyond a certain point, there is simply *too much*
intrinsic error in solving (2.7) for the "linear interpolation hypothesis" to be of any
use at level H3. Cf. §11.
([13]) on the *same* program
([14]) And were initially very perplexed by it. [Our initial *expectation* was that the
results would be *identical* -- or else very nearly so.]
([15]) Nor does one possess any kind of computational argument principle to help with
the bookkeeping [as with $\zeta(s)$] .

gap $12/R$, testing several independent z_j - batches, and by careful exercise of precautions (i)-(iv), especially (iii).

Appendix B contains an *example* of one of our codes. Some readers may be interested in seeing what the code actually looks like. (Due to space limitations, the associated "manual" has to be omitted.)

We now turn to the experimental results...

§5. The Even Eigenvalues Less Than 50.   In this experiment, we used the CRAY-1.
We considered 5 batches of z_j satisfying

$$A \overset{\le}{=} y_j \overset{\le}{=} 2 \quad \& \quad x_j = 0 \text{ or } \tfrac{1}{2} \quad \text{(i.e. } two \text{ vertical columns)}$$

with a "control" parameter $A \in (1, 1.100]$.   We then chose

$$H1 = .025 , \quad H2 = .001 , \quad H3 = 10^{-6}$$

in line with constraints (3.1)+(3.2).   Since the c_n "hump" was rather small [especially for $R < 25$] , we *omitted* the preliminary "weeding" of spurious R at level H2. We took $\gamma = \tfrac{1}{5}$ and used (4.2) as the filter.

Our results are displayed in tables 1 & 2.   In table 2, the *last* digit of each c_n may be off.   Compare: [7,appendix C] and [19] .

13.779751 (13,7)	35.502349 (23,12)	45.287438 (25,14)
17.738563 (15,8)	35.841677 (23,12)	45.361613 (25,14)
19.423481 (15,8)	36.677553 (23,12)	45.398470 (25,14)
21.315796 (17,8)	36.856349 (23,12)	46.101456 (25,14)
22.785909 (17,10)	37.825072 (23,13)	46.481402 (27,15)
24.112353 (17,10)	38.303276 (23,13)	46.653318 (27,15)
25.826244 (19,10)	39.168085 (23,13)	47.422896 (27,15)
26.152085 (19,10)	39.407532 (23,13)	47.926558 (27,15)
27.332708 (19,10)	39.773623 (23,13)	48.039331 (27,15)
28.530747 (19,10)	40.543351 (23,13)	48.741666 (27,15)
28.863394 (19,10)	40.688666 (23,13)	48.998308 (27,16)
30.410679 (21,11)	41.555578 (23,14)	49.683520 (27,16)
31.526582 (21,11)	41.883003 (23,14)	49.961697 (27,16)
31.566275 (21,11)	42.643489 (25,14)	50.089705 (27,16)
32.508118 (21,11)	42.922228 (25,14)	
32.891170 (21,11)	43.267182 (25,14)	
34.027884 (21,11)	44.077405 (25,14)	
34.456271 (23,11)	44.426348 (25,14)	

Even Eigenvalues for PSL(2,**Z**) .

The ordered pairs express (N, c_n "hump") .

TABLE   1

R	13.779751	17.738563	19.423481	21.315796	22.785909
c_2	1.549305	-.765456	-.692759	1.287534	.267693
c_3	.246900	-.977777	1.562349	1.251768	-.585496
c_5	.737	-1.015	-.03843	1.170	.03834
c_7	***	***	***	***	.99

R	24.112353	25.826244	26.152085	27.332708	28.530747
c_2	1.712436	.258066	-1.866163	-.209009	-1.460502
c_3	.881068	1.333742	-.403768	-.114727	.211383
c_5	-.35537	1.276361	-.160388	-.700599	1.432932
c_7	1.32	.743	-.619	-.252	1.1582

R	28.863394	30.410679	31.526582	31.566275	32.508118
c_2	.770445	1.346139	-.75896	.530999	1.812750
c_3	-1.559404	.186890	1.64391	-.847997	1.171151
c_5	.308238	1.350177	.84987	-1.392501	-.414880
c_7	-1.3132	.7649	.9736	-.54173	.323447

R	32.891170	34.027884	34.456271		
c_2	.173574	1.171189	-1.235721		
c_3	-.501554	-.947719	-.553993		
c_5	1.819350	-1.034348	.900853		
c_7	-.54885	1.14157	.61465		

Fourier Coefficients for the Even Eigenvalues Less than 35.

TABLE   2

We also include 1 slightly larger example (for use as a possible reference in other experiments).

R = 47.926558	
c_2 = .511058	c_7 = .028926
c_3 = 1.700681	c_{11} = -.295543
c_5 = -1.358583	c_{13} = -.08748

TABLE   3

To convey some of the accuracy, we remark that:

$$|c_4 - c_2{}^2 + 1| = 2 \times 10^{-8} \qquad |c_6 - c_2 c_3| = 4 \times 10^{-8}$$

$$|c_8 - c_2{}^3 + 2c_2| = 1.3 \times 10^{-7} \qquad |c_9 - c_3{}^2 + 1| = 1.3 \times 10^{-7}$$

$$|c_{10} - c_2 c_5| = 2.2 \times 10^{-7} \qquad |c_{12} - c_3 c_4| = 3.7 \times 10^{-6}$$

$$|c_{14} - c_2 c_7| = 5.3 \times 10^{-3} \quad \text{[for } N = 27 \text{ and a } c_n \text{ hump of about 15]} \quad .$$

Ramanujan-Petersson was checked for all R-values in table 1 -- at least out to the c_n "hump." There were *no* violations.

We should also mention that, once R reached 40 or so, certain R_n began to be "missed" on one batch or another. Cf. §4 paragraphs 6-7.

Sample runtimes for testing [the interval] $46 \lesssim R \lesssim 50$ are as follows:

$$\left\{\begin{array}{ll} \text{CRAY-1} & 414 \text{ sec.} \\ \text{XMP} & 315 \text{ sec.} \\ \text{CRAY-2 (cft)} & 293 \text{ sec.} \\ \text{CRAY-2 (cft77)} & 278 \text{ sec.} \end{array}\right\} \quad .$$

The anomaly mentioned at the end of §4 does <u>not</u> occur yet. [Incidentally: *each* Bessel function takes about .0019 / .0002 / .0002 seconds to compute on the CRAY-1 depending on whether $R > X$, $R \approx X$, $R < X$. The other machines are proportionately faster.]

§6. <u>The Odd Eigenvalues Less Than 50</u>. Here we used the same set-up as in §5 but only 4 batches instead of 5. Since the CRAY-1 was temporarily unavailable, we switched over to the CRAY-2.

9.533695	(11,6)	25.050855	(17,10)	32.932465	(21,12)
12.173008	(13,7)	26.056918	(19,10)	33.492331	(21,12)
14.358509	(13,7)	26.446996	(19,11)	33.570990	(21,12)
16.138073	(15,8)	27.284384	(19,10)	34.185970	(21,12)
16.644259	(15,8)	27.775921	(19,10)	34.695311	(23,12)
18.180918	(15,8)	28.510278	(19,10)	35.431665	(23,12)
19.484714	(15,8)	29.137588	(19,11)	35.666397	(23,12)
20.106695	(15,8)	29.546388	(19,10)	35.858674	(23,12)
21.479058	(17,9)	30.279049	(21,10)	36.331129	(23,12)
22.194674	(17,9)	30.404327	(21,10)	36.988815	(23,12)
23.201396	(17,9)	31.056534	(21,11)	37.295583	(23,12)
23.263712	(17,10)	31.916182	(21,11)	37.743925	(23,12)
24.419715	(17,10)	32.018406	(21,12)	38.120901	(23,12)

(continued)

38.442004 (23,12)	42.978654 (25,13)	47.178366 (27,14)
38.869607 (23,12)	43.385687 (25,13)	47.546230 (27,14)
39.432477 (23,12)	43.859382 (25,13)	47.823373 (27,14)
39.826868 (23,12)	44.282110 (25,13)	48.149810 (27,15)
40.272111 (23,12)	44.294967 (25,13)	48.355412 (27,15)
40.858127 (23,12)	44.777046 (25,13)	48.840152 (27,15)
40.880467 (23,12)	45.112201 (25,14)	48.896682 (27,15)
40.990437 (23,13)	45.686380 (25,14)	49.105724 (27,15)
41.754473 (23,13)	45.782821 (25,14)	49.439178 (27,16)
42.152733 (23,13)	45.954420 (25,14)	49.991221 (27,15)
42.485562 (25,13)	46.566346 (27,14)	
42.646363 (25,13)	46.839220 (27,14)	

Odd Eigenvalues for $PSL(2,\mathbf{Z})$.

The ordered pairs express $(N, c_n\text{"hump"})$.

TABLE 4

R	9.533695	12.173008	14.358509	16.138073	16.644259
c_2	−1.06833	.289252	−.230915	1.161855	−1.540228
c_3	−.4563	−1.201858	.69560	−1.281972	.977493
c_5	***	.042	−1.28	−.756	−.105
c_7	***	***	***	***	***

R	18.180918	19.484714	20.106695	21.479058	22.194674
c_2	.374063	−1.700188	.858848	−.656250	1.596844
c_3	.101958	−.614565	.187279	.226442	−1.116480
c_5	.6372	.8199	−1.3956	1.08229	−.63825
c_7	***	***	***	.423	−1.01

(continued)

R	23.201396	23.263712	24.419715	25.050855	26.056918
c_2	.169949	−1.447094	.965541	−1.053870	1.159119
c_3	1.493056	−1.536666	−.690260	.552022	.598888
c_5	−.93998	.10696	1.315804	−.73366	−1.089323
c_7	−.57	.62	−.545	1.546	−1.2786

R	26.446996	27.284384	27.775921	28.510278	29.137588
c_2	−.637458	−1.20563	.948347	−1.314095	−.085103
c_3	−1.358607	1.66456	−.192092	−1.410043	.820706
c_5	1.382687	−.45130	.164078	−1.325850	1.031232
c_7	−.00879	−.80	1.131	.0840	−1.0347

R	29.546388	30.279049	30.404327	31.056534	31.916182
c_2	1.723163	−1.781785	−1.685405	.860109	−.963673
c_3	−.395634	1.005447	−1.262392	1.161835	−.168495
c_5	−1.471689	−.794101	.420029	−.165096	−.030422
c_7	−.1287	−.1079	1.7351	−1.1332	−.33567

R	32.018406	32.932465	33.492331	33.570990	34.185970
c_2	1.607004	.601183	−1.71021	−.576907	1.507063
c_3	−1.561208	.755352	1.08331	−1.861226	−.116023
c_5	.641005	−.238555	1.06151	−.552746	.363646
c_7	.03243	1.02183	−.8569	−.46703	−1.79833

R	34.695311				
c_2	−.332079				
c_3	.548638				
c_5	−.065240				
c_7	1.03357				

Fourier Coefficients for the Odd Eigenvalues Less than 35.

TABLE 5

We again include 1 slightly larger example.

R = 47.178366	
c_2 = 1.314569	c_7 = .397455
c_3 = .522603	c_{11} = .44840
c_5 = -.189176	c_{13} = -1.724

TABLE   6

Ramanujan-Petersson was checked in table 4, at least out to the c_n "hump." There were *no* violations.

§7. Even Eigenvalues Around R = 125 . We did not know what to expect here. We decided to run a batch of experiments to see if we could find the 15 or so expected even eigenvalues in the interval [124.875,126.325] . The interest centered on what sort of accuracy could be obtained (using *just* single precision), and on *how many* "trials" would have to be run. In general: we took

$$H1 = .025 \quad , \quad H2 = .001 \quad , \quad H3 = 10^{-6}$$

and used (4.2) with $\gamma = \frac{1}{3}$. Note that H1 and H2 are consistent with constraints (3.1) + (3.2).

The z_j - batches were similar to §5 except that we also used some having 3 or 4 vertical columns.

In each case: the relevant N is about 55 and the c_n "hump" is about 32.

There were 26 jobs in all.

Here is a summary of our results. **[** We *believe* that this listing is complete. **]**

R	appears in	out of (jobs)
124.898691	13	21
124.994438	7	21
125.036859	12	21
125.313840	14	21
125.347558	12	21
125.523988	20	21
125.673602	16	21
125.896473	18	21

(continued)

126.018778	7	21
126.066382	9	21
126.113995	21	21
126.250406	5	5
126.313569	3	5
126.321149	1	5

Even Eigenvalues in [124.875 , 126.325]

Average Gap = .0960

TABLE 7

21 jobs dealt with [124.875,126.125] ; 5 jobs dealt with [126.125,126.325] . The frequency count shows that some R_n are (apparently) more "visible" than others. The following statistics may also be of interest:

completeness probabilities for [124.875,126.125]

jobs with 2 columns averaged 8.09 (true values) out of 11   (over 11 jobs)

jobs with 3 columns averaged 6.67 (true values) out of 11   (over 6 jobs)

jobs with 4 columns averaged 5.00 (true values) out of 11   (over 4 jobs)

[the corresponding ratios are .74, .61, .45]

average runtimes for [124.875,126.125]

17 jobs on CRAY2 : 791 sec. per job

4 jobs on XMP :   695 sec. per job

[the "noisier" jobs were all run on the CRAY-2]   .

In most jobs: *some* spurious R-values managed to slip through all the way til the end. In the "noisiest" jobs, this number occasionally ran as high as 70% of the final listing. *True* R-values typically have their first few "Hecke differences" *vanishing* to between 4 and 7 decimal places.

For the sake of completeness, we include a coefficient listing for 3 of our better R-values. As usual: the last digit in any c_n may be off.

R	125.313840	125.347558	125.523988
c_2	−.332696	.523115	−1.530691
c_3	1.056574	−1.012324	.707551
c_5	−.946097	.586322	1.222769
c_7	.324132	.225425	.706786
c_{11}	.040097	.106276	−1.132417
c_{13}	.479454	−1.206671	−.496056
c_{17}	1.254990	.602756	.056324
c_{19}	.893140	−.231553	1.094387
c_{23}	1.605562	.407487	−.725439
c_{29}	.85671	−1.25613	−1.67296
c_{31}	−.25088	−.84230	1.02462

TABLE 8

To convey some of the accuracy, we remark that:

$$|c_{10} - c_5 c_2| = \begin{cases} 2.4 \times 10^{-7} & \text{for} \quad 125.313840 \\ 7.4 \times 10^{-7} & \text{for} \quad 125.347558 \\ 9.4 \times 10^{-7} & \text{for} \quad 125.523988 \end{cases}$$

$$|c_{20} - c_5 c_4| = \begin{cases} 9.8 \times 10^{-7} \\ 2.0 \times 10^{-6} \\ 1.6 \times 10^{-6} \end{cases}$$

$$|c_{24} - c_8 c_3| = \begin{cases} 1.0 \times 10^{-6} \\ 4.8 \times 10^{-7} \\ 1.6 \times 10^{-6} \end{cases}$$

$$|c_{30} - c_6 c_5| = \begin{cases} 4.6 \times 10^{-6} & \text{for} \quad 125.313840 \\ 1.1 \times 10^{-5} & \text{for} \quad 125.347558 \\ 1.5 \times 10^{-5} & \text{for} \quad 125.523988 \end{cases}$$

Ramanujan-Petersson was checked for all 14 R-values in table 7 (at least out to the c_n "hump"). No violations were found. The "extremal" values are:

$$\left\{ \begin{array}{lll} c_5 = 1.81457 & \text{for} & R = 126.018778 \\ c_{29} = -1.95662 & \text{for} & R = 126.250406 \end{array} \right\} \quad .$$

§8. Even Eigenvalues Around R = 250 . Here we used pretty much the same set-up as in §7, but did a bit of "tampering" with γ and H3 . In general, we took

$$H1 = .050 \quad , \quad H2 = .001 \quad , \quad H3 = 10^{-6} \quad , \quad \gamma = \frac{1}{10} \quad .$$

The "$\frac{1}{10}$" seemed to work fine except in a few cases. There we went to a finer and/or wider filtering grid; i.e.

$$H3 = \frac{1}{4} \times 10^{-6} \quad \text{and/or} \quad \gamma = \frac{1}{2} \quad .$$

We also experimented with using LINPACK to solve (2.7) -- in place of calling our own (*unoptimized*) subroutine SMAT.

There were 23 full-scale jobs plus a *similar* number of "shortcut" versions. Of the full-scale jobs: 21 had 2 columns, while 2 had 3 columns.

In each case: the relevant N was about 100 , while the c_n "hump" was about 68.

Here is a summary of our results. [We *believe* this listing to be complete.] The R interval is $250 \lesssim R \lesssim 250.546$; the expected number of eigenvalues is 11 .

R	appears in	out of (jobs)
250.014292	14	23
250.157023	18	23
250.171541	9	23
250.199849	11	23
250.220519	8	23
250.294156	4	23
250.323063	15	23
250.360005	5	23
250.512602	2	23
250.521575	3	23

Even Eigenvalues in [250,250.546]

Average Gap = .0480

TABLE 9

The following statistics apply:

<u>completeness probabilities for [250,250.546]</u>

jobs with 2 columns averaged 4.14 (true values) out of 10   (over 21 jobs)

jobs with 3 columns averaged 1.00 (true values) out of 10   (over 2 jobs)

<u>average runtimes for [250,250.546]</u>

CRAY2 jobs with SMAT :	726 sec.	(8 jobs)
CRAY2 jobs with LINPACK :	734 sec.	(6 jobs)
XMP jobs with LINPACK :	789 sec.	(4 jobs)
finer and/or wider CRAY2 jobs with LINPACK:	1646 sec.	(5 jobs)
standard "short-cut" job on either machine:	200 seconds	

<u>typical runtimes for solving (2.7) with N = 100</u>

CRAY2 with LINPACK	.006 sec
CRAY2 with SMAT	.086 sec

[though slower, SMAT occasionally gave *better* answers than LINPACK]   •

For R $\sim$ 250 , *true* R-values typically have their first few "Hecke differences" vanishing to between 4 and 6 decimal places.

As in §7 we include a coefficient listing for 3 of our better R-values.

R	250.014292	250.157023	250.171541
c_2	1.963398	.934272	-.847139
c_3	.966543	.555460	.626670
c_5	-.572737	1.400902	-1.558957
c_7	-1.908284	-.617719	-.031878
c_{11}	.258535	.129422	-.455558
c_{13}	.485796	-.875756	-.388115
c_{17}	.148212	-.355089	-1.468538
c_{19}	-.748242	-1.216197	-.211623
c_{23}	-.576633	-.732391	.234726
c_{29}	-.542465	-.625467	1.360191
c_{31}	-.917826	-.330008	-1.85193
c_{37}	-.561364	-.462475	1.24353
c_{41}	1.330662	-.902229	1.43213
c_{43}	1.159073	-.503389	.43388

(continued)

c_{47}	.188167	.923052	1.60671
c_{53}	−.372197	.499237	1.01753
c_{59}	1.832051	−1.539013	−.23515
c_{61}	−.27625	.43774	−.21213
c_{67}	.5898	−.864	.0334

$$\boxed{\text{TABLE } 10}$$

To convey some of the accuracy, we remark that:

$$|c_{10} - c_5 c_2| = \begin{cases} 2.0 \times 10^{-8} & \text{for} & 250.014292 \\ 5.4 \times 10^{-6} & \text{for} & 250.157023 \\ 6.6 \times 10^{-6} & \text{for} & 250.171541 \end{cases}$$

$$|c_{30} - c_5 c_3 c_2| = \begin{cases} 2.4 \times 10^{-7} \\ 5.4 \times 10^{-6} \\ 1.5 \times 10^{-5} \end{cases}$$

$$|c_{60} - c_5 c_4 c_3| = \begin{cases} 2.1 \times 10^{-6} \\ 6.1 \times 10^{-6} \\ 5.4 \times 10^{-6} \end{cases}$$

$$|c_{65} - c_{13} c_5| = \begin{cases} 1.6 \times 10^{-4} & \text{for} & 250.014292 \\ 9.3 \times 10^{-4} & \text{for} & 250.157023 \\ 2.5 \times 10^{-4} & \text{for} & 250.171541 \end{cases} \quad \bullet$$

Ramanujan-Petersson was checked for all 10 values in table 9 (at least out to the c_n "hump"). No violations were found. The extremal values were:

$$\begin{cases} c_2 = 1.96340 & \text{for} & R = 250.014292 \\ c_7 = -1.90828 & \text{for} & R = 250.014292 \end{cases} \quad \bullet$$

§9. <u>Even Eigenvalues Around $R = 500$</u>. The last "full-scale" search was made at 500. As in §8 we did a bit of tampering with γ and H3. In general, we took:

$$H1 = .050 \quad , \quad H2 = .0005 \quad , \quad (H3, \gamma) = (10^{-6}, \tfrac{1}{10}) \text{ or } (\tfrac{1}{2} \times 10^{-6}, \tfrac{1}{5}) \quad .$$

In several cases, however, we went with

$$H3 = 1 \times 10^{-7} \quad , \quad 2.5 \times 10^{-7} \quad , \quad \text{and/or} \quad \gamma = \frac{1}{2}$$

to try to obtain greater accuracy.

There were 16 full-scale jobs plus a somewhat *larger* number of "short-cut" versions. All the jobs had 2 columns.

In each case: the relevant N was about 188 , while the c_n "hump" was about 137.

Here is a summary of our results. **[** We *believe* this listing to be complete.**]** The chosen interval was $500 \overset{\leq}{=} R \overset{\leq}{=} 500.298$; the expected number of eigenvalues is 12 .

R	appears in	out of (jobs)
500.038825	8	16
500.048046	2	16
500.066461	14	16
500.075997	14	16
500.113941	9	16
500.138065	3	16
500.141707	2	16
500.151972	4	16
500.214756	3	16
500.232299	3	16
500.271235	1	16
500.283551	13	16

Even Eigenvalues in [500,500.298]

Average Gap = .0240

TABLE   11

The following statistics apply:

completeness probabilities for [500,500.298]

jobs with 2 columns averaged 4.75 (true values) out of 12   (over 16 jobs)

[*or:* 40%]

average runtimes for [500,500.298]		
CRAY-2 jobs with SMAT:	3981 sec	(3 jobs)
CRAY-2 jobs with LINPACK:	2215 sec	(4 jobs)
XMP jobs with LINPACK:	2597 sec	(4 jobs)
finer and/or wider jobs, either machine:	4709 sec	(5 jobs)
standard "short-cut" job on XMP, with LINPACK:	612 sec	
standard "short-cut" job on CRAY-2, with LINPACK:	460 sec	
standard "short-cut" job on CRAY-2, with SMAT:	730 sec	

typical runtimes for solving (2.7) with $N \sim 200$	
CRAY-2 with LINPACK	.030 sec
CRAY-2 with SMAT	.621 sec

For $R \sim 500$, true R-values typically have their first few "Hecke differences" vanishing to between 4 and 5 decimal places. If one goes out to $n = 50$, the number of places may *drop* to 3.

The number of *spurious* values slipping through til the very end may run as high as 50 or 60% (of the final listing) in the "noisiest" jobs.

Here are the coefficients for 2 of our better R-values.

R	500.066461	500.283551
c_2	−.71386	1.10979
c_3	.82261	−.65140
c_5	.26012	−1.33193
c_7	1.48781	.95336
c_{11}	.25535	−.29227
c_{13}	.80536	−.84352
c_{17}	−.23060	−1.24396
c_{19}	−1.47045	−1.36471
c_{23}	.58742	−.21683
c_{29}	−1.52386	−.35004
c_{31}	−1.80923	1.55889
c_{37}	.81740	1.57655
c_{41}	−1.29589	−.11597
c_{43}	.26127	1.35647
c_{47}	−.87901	−1.59240
c_{53}	−.64300	−1.54331
c_{59}	−.15085	−.09993
c_{61}	−.14501	1.53694

(continued)

c_{67}	-.94672	-.98225
c_{71}	-.35015	-.57396
c_{73}	-.50796	1.18253
c_{79}	-1.73634	1.59923
c_{83}	-.41151	-.54689
c_{89}	-.91326	-1.54487
c_{97}	1.60887	.22588
c_{101}	1.78973	1.05331
c_{103}	-1.63371	1.50197
c_{107}	-.59332	.89831
c_{109}	-.39324	.47863
c_{113}	.71463	.90804
c_{127}	1.82748	.87385
c_{131}	1.1836	.5238

$$\boxed{\text{TABLE} \quad 12}$$

To convey some of the accuracy, we remark that

$$|c_{10} - c_5 c_2| = \begin{cases} 8.8 \times 10^{-6} & \text{for} \quad 500.066461 \\ 1.2 \times 10^{-5} & \text{for} \quad 500.283551 \end{cases}$$

$$|c_{30} - c_5 c_3 c_2| = \begin{cases} 8.8 \times 10^{-6} \\ 1.1 \times 10^{-6} \end{cases}$$

$$|c_{60} - c_5 c_4 c_3| = \begin{cases} 7.5 \times 10^{-6} \\ 4.7 \times 10^{-6} \end{cases}$$

$$|c_{100} - c_{25} c_4| = \begin{cases} 4.3 \times 10^{-6} \\ 4.6 \times 10^{-6} \end{cases}$$

$$|c_{126} - c_{14} c_9| = \begin{cases} 2.5 \times 10^{-5} \\ 6.8 \times 10^{-6} \end{cases}$$

$$|c_{130} - c_{65} c_2| = \begin{cases} 6.3 \times 10^{-4} & \text{for} \quad 500.066461 \\ 3.1 \times 10^{-4} & \text{for} \quad 500.283551 \end{cases} \qquad \bullet$$

Ramanujan-Petersson was checked for all 12 values in table 11 (at least out to the c_n "hump"). No violations were found. The extremal values were:

$$\begin{cases} c_2 = 1.89788 & \text{for} \quad R = 500.151972 \\ c_2 = -1.94412 & \text{for} \quad R = 500.232299 \end{cases} \qquad \bullet$$

§10. <u>Gaining Greater Accuracy</u>.   In sections 5 and 6, we simply *took* H3 = 10^{-6}.
As a *measure* of the intrinsic accuracy of the Bessel function (and linear algebra)
routines, it is very tempting to substitute smaller & smaller values for H3 to see
exactly *how much* accuracy can *ultimately* be obtained (in single precision).

"Proximity to the truth" is measured by precaution (iii) in §2.

To save time, we used a "bare-bones" form of the program which contained *only one*
level (namely H3). We used the CRAY2 and looked at successively smaller neighborhoods
of the (even) eigenvalues:

$$13.779751^+ \qquad \text{and} \qquad 41.555578^+ \qquad \bullet$$

In the case of 13.779751, we experimented with several different z_j - batches (of 2
vertical columns) *and* several different N .   The best results were obtained by taking
N to be 13 [which corresponds to (only) 10^{-9} in (2.6)].

We found that:

R = 13.779751351890$^+$
with
c_2 = 1.54930447794
c_3 = .24689977245
c_5 = .737060383
c_7 = $-$.261421

and

$$\left| c_4 - c_2^2 + 1 \right| = 1 \times 10^{-10} \quad , \qquad \left| c_6 - c_2 c_3 \right| = 4.8 \times 10^{-8} \quad ,$$

$$\left| c_8 - c_2^3 + 2c_2 \right| = 1.8 \times 10^{-5} \quad , \qquad \left| c_9 - c_3^2 + 1 \right| = 2.0 \times 10^{-3} \quad ,$$

$$\left| c_{10} - c_2 c_5 \right| = 5.5 \times 10^{-2} \qquad \bullet$$

The c_n "hump" has (thus) been "pushed" all the way up to N$-$2 or N$-$1 [from 7].

The other *batches* gave either 1351889$^+$ or 1351890$^+$, but their Hecke differences
(at n = 8,9,10) were significantly larger...

The foregoing result compares *very* favorably with [7,appendix C] and [19] .

For R = 41.555578, we basically looked at *just one* z_j - batch but *varied* N between
23 $\sim$ 27 .   The results were pretty much the same in each case.   We obtained:

$$R = 41.5555776736^+$$

with
$c_2 = 1.0589646368$
$c_3 = -.9589822875$
$c_5 = -.6334749873$
$c_7 = -.8135527894$
$c_{11} = 1.1360708$
$c_{13} = -1.11568$

and (for $N = 23$)

$$E_4 = 3 \times 10^{-10} \qquad E_6 = 1 \times 10^{-10} \qquad E_8 = 1 \times 10^{-10}$$

$$E_9 = 1.7 \times 10^{-9} \qquad E_{10} = 2.7 \times 10^{-8} \qquad E_{12} = 1.2 \times 10^{-5}$$

$$E_{14} = 5.9 \times 10^{-3} \qquad \text{[in an obvious notation]} \qquad .$$

Here the c_n "hump" has been pushed from 14 out to 16 or so.

The results of these (two) "refinements" are *perfectly consistent* with an overall accuracy of $11 \sim 12$ significant figures in KBESS. Cf. §3 near (3.2).

§11. Concluding Remarks. The results of these experiments ([16]) are, from one standpoint, very promising. They show that the "K-Bessel barrier" is now *all* but removed -- and that the real difficulties (*mathematically*) are more in line with linear algebra. Concerning the latter: we have seen that, by employing a suitable mechanism to filter out c_k noise, it is actually possible to get *quite* far using ONLY single precision.

There are, however, *several* features that seem a bit disturbing here -- especially in connection with any kind of *more systematic* (computational) analysis. Specifically:

(a) the cpu time is starting to become a bit large (as $R \nearrow$) ;

(b) because of the [inevitable] regions of static, there is an increasing need to concern oneself with the notion of "completeness probability" in any particular job ;

(c) the *anomaly* mentioned at the end of §4 becomes more & more noticeable as $R \nearrow$;

(d) we do not possess any kind of *apriori technique* for determining the number of λ_n actually present in a given interval (A,B) .

One is *not* really in a position to do too much about item (d) right now. [Similar difficulties occur in computing the zeros of many other (arithmetically significant) functions. Compare ([15]) in §4.]

([16]) *viz.* §§5-10

With regard to (a) and (b), one is *inclined* to say that we have now reached a kind of "intermediate" stage, where the *next* step should (effectively) be one "last, massive, systematic optimization."

Testing *several* z_j - batches seems inescapable -- at least if one wants to use only single precision. For this reason: it is tempting to suggest that the z_j - batches should all be treated *in parallel* at levels H1 & H2 -- and that any "passage" to level H3 should occur *only* in those cases judged to be most stable.

Thus far, we have been content to use local transversality of $c_4 + 1 - c_2^2$ [at the zeros of $c_4 + 1 - c_2^2$] as the basic "triggering" mechanism. This idea worked reasonably well out to $R = 500$ but there might be better ways...

This matter becomes important in connection with $(^{12})$ in §4. In a parallel program of the type suggested above, one would naturally like to use the "short-cut" method as far as possible. The *problem* is that, beyond a certain point, one simply *has* to expect that the H2 - level data will be plagued by sufficient machine noise to make any kind of linear interpolation down to level H3 virtually useless. Our preliminary experiments suggest that this point is already reached at $R = 1000$.

In a final series of experiments, we used the CRAY2 to do "a bit of exploration" around $R = 1000$ (with $N \approx 360$). Our *shortcut* jobs were *unable* to come up with even a *single* serious candidate for a true R_n . To make things worse, the "full-scale" jobs that we then ran (as a last resort) also failed!! The fact that H3 was 10^{-7} (instead of 10^{-6}) did *not* seem to make much difference. It appeared that the "triggering" mechanism for the filter was simply *unable* to identify *any* correct H2 - intervals.

Some explanation for this failure is already "visible" in the "velocity" fluctuations for the c_k at level H2 . The striking thing *at* $R = 1000$ is that these fluctuations are typically 100 times as large as the corresponding fluctuations at $R = 500$ (*even for* $k = 2$). The larger these numbers, the greater the "noise" at level H2. $(^{17})$

To treat $R \gtreqless 1000$, it seems *essential* to either:

(A) develop a better triggering mechanism ; or

(B) somehow *reduce* the c_k noise-level at H2 .

One hopeful point is this. The noise levels in the various c_k are NOT uncorrelated. There is a kind of "uniformity" with respect to k (at least when Gauss elimination is used). This uniformity partly explains why the filter itself is able to function. It seems reasonable to expect that some use can be made of this fact.

On the over-all matter of further optimization, we simply mention the following points:

(1) perhaps $\exp(\frac{\pi}{2}R)K_{iR}(X)$ can be satisfactorily computed using a *wider grid* in the numerical integration ; [this would certainly speed things up] $(^{18})$

$(^{17})$ Values like 10^5 are *quite* common for H2 $= 1.25 \times 10^{-4}$ (say) and $k \lesssim 6$.

$(^{18})$ Iterative techniques are *another* possibility. Cf. [10,p.1369] , however.

(2) one should *also* experiment with using the *largest* possible H1 & H2 values ;

(3) perhaps there are better methods of solving (2.7) than by standard Gauss elimination ;

(4) there's an obvious need for some better "triggering" mechanisms ;

(5) it would be highly advantageous to let the *machine* decide in the final listing (for each run) *which* of the R-values is "true" and which is "spurious" (possibly by referring to the results from earlier runs) ;

(6) with regard to parallelism (and *new types* of triggering mechanisms), *note* that any "true" R must be invariant under the process of taking arbitrary linear combinations of (2.7) with respect to distinct z_j - batches.

In conclusion: it virtually goes without saying that one *expects* that Stark's method [19] can be combined with our techniques to yield *many many* more c_k for each R_n . Stark's use of iteration in solving for c_k seems particularly suggestive. Cf. (3) above. ⟨See also ref. [52] on page 105.⟩

We hope to report on further developments in these areas in the near future.

REFERENCES

1. N.L.Balazs and A.Voros, Chaos on the pseudosphere, Phys. Reports 143(3)(1986) 109-240.

2. M.Berry, Quantum chaology, Proc. Royal Soc. London A413(1987) 183-198. See also: Proc. Royal Soc. London A400(1985) 229-251.

3. O.Bohigas, M.J.Giannoni, and Ch. Schmit, Spectral fluctuations, random matrix theories, and chaotic motion, Springer Lecture Notes in Physics 262(1986) 118-138.

4. P.Cartier, Some numerical computations relating to automorphic functions, in *Computers in Number Theory* (ed. by A.O.L.Atkin and B.J.Birch), Academic Press,1971, pp.37-48.

5. G.Golub and C.Van Loan, *Matrix Computations*, Johns Hopkins Univ. Press, 1983 , especially pages 25-27 and 71-72.

6. D.A.Hejhal, The Selberg trace formula and the Riemann zeta function, Duke Math. J. 43(1976) 441-482.

7. D.A.Hejhal, *The Selberg Trace Formula for PSL(2,R)* , volume 2, Springer Lecture Notes 1001(1983).

8. D.A.Hejhal, Some observations concerning eigenvalues of the Laplacian and Dirichlet L-series, in *Recent Progress in Analytic Number Theory* (ed. by H.Halberstam and C.Hooley), volume 2, Academic Press, 1981, pp.95-110.

9. D.A.Hejhal and E.Bombieri, Sur les zéros des fonctions zêta d'Epstein, Comptes Rendus Acad. Sci. Paris 304(1987) 213-217.

10. D.A.Hejhal, Zeros of Epstein zeta functions and supercomputers, in *Proceedings of the International Congress of Mathematicians*, Berkeley, 1986, pp.1362-1384.

11. D.A.Hejhal, Some remarks about cusp forms: holomorphic and non-holomorphic, Technical Report No. 1984-26, Chalmers Univ. of Tech. (Sweden), 1984, 33pp .

12. D.A.Hejhal and B.Berg, Some new results concerning eigenvalues of the non-Euclidean Laplacian for PSL(2,**Z**), Technical Report No. 82-172, University of Minnesota, 1982, 7 pp .

13. H.Iwaniec, Non-holomorphic modular forms and their applications, in *Modular Forms* (ed. by R.A.Rankin), Ellis-Horwood Ltd., 1984, pp.157-196.

14. N.V.Kuznecov, Petersson's conjecture for cusp forms of weight zero and Linnik's conjecture; sums of Kloosterman sums, Math. USSR Sbornik 39(1981) 299-342.

15. A.M.Odlyzko, On the distribution of spacings between zeros of the zeta function, Math. of Comp. 48(1987) 273-308.

16. G.Pólya, Bemerkung über die Integraldarstellung der Riemannschen ξ - Funktion, Acta Math. 48(1926) 305-317.

17. A.Selberg, Harmonic analysis and discontinuous groups in weakly symmetric Riemannian spaces with applications to Dirichlet series, J. Indian Math. Soc. 20(1956) 47-87.

18. C.Series, Some geometrical models of chaotic dynamics, Proc. Royal Soc. London A 413(1987) 171-182.

19. H.Stark, Fourier coefficients of Maass waveforms, in *Modular Forms* (ed. by R.A. Rankin), Ellis-Horwood Ltd., 1984, pp.263-269.

20. A.B.Venkov, *Spectral Theory of Automorphic Functions*, Proc. Steklov Inst. of Math. 153(1982). (English Translation)

21. G.N.Watson, *A Treatise on the Theory of Bessel Functions*, 2nd edition, Cambridge Univ. Press, 1944.

22. M.Wilkinson, Random matrix theory in semiclassical quantum mechanics of chaotic systems, J. Phys. A: Math. Gen. 21(1988) 1173-1190.

23. A.Winkler, Cusp forms and Hecke groups, J. Reine Angew. Math. 386(1988) 187-204.

APPENDIX B

```
      PROGRAM DHU30A
C     experimental eigenvalue program
C     D.A.HEJHAL // October 1988
C     CRAY2 VERSION -- SINGLE PRECISION
C     this program uses an adjustable Lagrange interpolation to
C     reduce the number of KBMAT calls;
C     it also uses TRIM indices (lifted from SCALE.f);
C     and exploits ------
C     weighted Hecke indices
C     transversality controls
C     grid levels H3,H4,H5
C     full KBMAT calls at H3
C     full SMAT calls at H4
C     full SMAT calls for MAG2 -- near "transverse" R values --
C     to improve c(k) accuracy in NON-NOISY cases (cf. control #)
C     flexible program segments
C     control indices for noise/error/distortion
      IMPLICIT REAL(A-H,P-Z)
      PARAMETER(MAXDEG=51)
      PARAMETER(MG=16)
      PARAMETER(NMAX=60,NPAX=60)
      PARAMETER(NTRIM=1)
      PARAMETER(IPRINT=1)
      PARAMETER(LONG=60)
      PARAMETER(MAG=25)
      PARAMETER(LONGER=LONG*MAG)
      PARAMETER(MAG2=1000)
      PARAMETER(MTW=9+2*MAG2)
      PARAMETER(NDEG=11)
C     make certain that NDEG is odd!!
      DIMENSION XA(NMAX),YA(NMAX),XP(NMAX),YP(NMAX)
      DIMENSION R3(0:LONG)
      DIMENSION R4(0:LONGER)
      DIMENSION RZ(LONGER,NTRIM), RZT(LONGER,NTRIM)
      DIMENSION ITRIM(NTRIM), LCOUNT(NTRIM)
      DIMENSION LKOUNT(NTRIM)
      DIMENSION GIANT(NMAX,NPAX,0:NDEG)
      DIMENSION U(NMAX,NPAX)
      DIMENSION UTMP1(NMAX,NPAX), UTMP2(NMAX,NPAX)
      DIMENSION CX1(NMAX)
      DIMENSION CDH(0:LONGER,NTRIM,NMAX)
      DIMENSION CDHH(0:MTW,NMAX), WH(MTW)
      DIMENSION R97(0:MTW)
      DIMENSION CDZ(LONGER,NTRIM,NMAX)
      DIMENSION BU(NMAX)
      DIMENSION CU(-2:3,NMAX)
      DIMENSION VE(-1:1,NTRIM,NMAX)
      DIMENSION VDZ(LONGER,NTRIM,NMAX)
      DIMENSION R4T(LONGER,NTRIM), R6T(LONGER,NTRIM)
      DIMENSION VDX(LONGER,NTRIM,NMAX)
      DIMENSION VDY(LONGER,NTRIM,NMAX)
      DIMENSION A1(16),W(16)
      DIMENSION AD(MG),WD(MG)
      INTEGER*8 FAC(20)
      DIMENSION RF(20),RG(20)
      COMMON /DENNIS/ AD,WD,PI,RTM,PIH
      COMMON /DH2/ RF,RG
      COMMON /DH3/ XA,YA,XP,YP
      COMMON /DH4/ NN1R,NN1L,NN2,NN3,PI2
      COMMON /DAH/ GIANT
```

```
C
C
      UDEG-FLOAT(NDEG)
      NA97-INT((.5EO)*(UDEG-1.0EO)+(1.0E-4))
C
C
      FAC(1)-1
      DO 10  J-2,20
         FAC(J)-J*FAC(J-1)
 10   CONTINUE
      DO 12  J1-1,20
         RF(J1)-(1.0EO)/FLOAT(FAC(J1))
         RG(J1)-(1.0EO)/FLOAT(J1)
 12   CONTINUE
      PI-(4.0EO)*ATAN(1.0EO)
      PIH-(.5EO)*PI
      PI2-(2.0EO)*PI
      A1(1)- .98940 09349 91650 EO
      A1(2)- .94457 50230 73233 EO
      A1(3)- .86563 12023 87832 EO
      A1(4)- .75540 44083 55003 EO
      A1(5)- .61787 62444 02644 EO
      A1(6)- .45801 67776 57227 EO
      A1(7)- .28160 35507 79259 EO
      A1(8)- .09501 25098 376374 EO
      A1(9)--A1(8)
      A1(10)--A1(7)
      A1(11)--A1(6)
      A1(12)--A1(5)
      A1(13)--A1(4)
      A1(14)--A1(3)
      A1(15)--A1(2)
      A1(16)--A1(1)
      W(1)- .02715 24594 117541 EO
      W(2)- .06225 35239 386479 EO
      W(3)- .09515 85116 824928 EO
      W(4)- .12462 89712 55534 EO
      W(5)- .14959 59888 16577 EO
      W(6)- .16915 65193 95003 EO
      W(7)- .18260 34150 44924 EO
      W(8)- .18945 06104 55068 EO
      W(9)-W(8)
      W(10)-W(7)
      W(11)-W(6)
      W(12)-W(5)
      W(13)-W(4)
      W(14)-W(3)
      W(15)-W(2)
      W(16)-W(1)
      MM-MG/16
      RTM-(1.0EO)/FLOAT(2*MM)
      DO 23  J3-0,MM-1
      DO 24  K3-1,16
         N3-16*J3
         WD(N3+K3)-W(K3)
         AD(N3+K3)-FLOAT(1+(2*J3))-A1(K3)
 24   CONTINUE
 23   CONTINUE
C
C
C     INPUT DATA:
C     in addition to PARAMETERS at top
C
      R1-124.75EO
      R2-126.25EO
      NN1R-4
      NN1L-16
```

```
            NN2=4
            NN3=4
            DATA ITRIM /55/
            NR=30
            NC=2
            DO 3 J=1,NR
            DO 4 I=1,NC
              XA(I+(J-1)*NC)=(.5E0)*FLOAT(I-1)/FLOAT(NC-1)
              YA(I+(J-1)*NC)=1.10E0+(FLOAT(J)*(.9E0)/FLOAT(NR))
      4     CONTINUE
      3     CONTINUE
      C
      C     END-OF-DATA
      C
            H3=(R2-R1)/FLOAT(LONG)
            H4=H3/FLOAT(MAG)
            H5=H4/FLOAT(MAG2)
            DO 5102 JJ=0,LONG
              R3(JJ)=R1+H3*FLOAT(JJ)
      5102  CONTINUE
            DO 5103 K=1,NMAX
              S=(XA(K)**2)+(YA(K)**2)
              XP(K)=-XA(K)/S
              YP(K)=YA(K)/S
      5103  CONTINUE
            DO 5104 JJ=0,LONGER
              R4(JJ)=R1+H4*FLOAT(JJ)
      5104  CONTINUE
      C
      C
      C
            DO 5301 JX=0,NDEG
            R=R3(JX)
            CALL KBMAT(R,U)
            DO 5302 KO=1,NMAX
            DO 5303 K=1,NMAX
              GIANT(K,KO,JX)=U(K,KO)
      5303  CONTINUE
      5302  CONTINUE
      5301  CONTINUE
      C
      C     giant do-loop follows:
      C
            DO 74 J=0,LONG-NDEG
      C
      C
            IF(J.NE.0) THEN
            RR97=R3(J+NDEG)
            CALL KBMAT(RR97,U)
CDIR$       NEXTSCALAR
            DO 5401 I=0,NDEG-1
            DO 5402 KO=1,NMAX
            DO 5403 K=1,NMAX
              S=GIANT(K,KO,I+1)
              GIANT(K,KO,I)=S
      5403  CONTINUE
      5402  CONTINUE
      5401  CONTINUE
            DO 5502 KO=1,NMAX
            DO 5503 K=1,NMAX
              S=U(K,KO)
              GIANT(K,KO,NDEG)=S
      5503  CONTINUE
      5502  CONTINUE
            ENDIF
      C
      C
```

```
            DO 6101 JB=0,MAG
            SX=FLOAT(NA97)+FLOAT(JB)/FLOAT(MAG)
            CALL FLAG(SX,UTMP1)
            M=JB+(J+NA97)*MAG
            DO 6102 I4=1,NTRIM
C           prepare an adjusted copy of UTMP1
C           note that the throw-away row can be modified!!!!!
            NMAT=ITRIM(I4)-1
            DO 6123 K=1,NMAT
              UTMP2(K,NMAT+1)=-UTMP1(K,1)
    6123    CONTINUE
            DO 6124 K0=1,NMAT
            DO 6125 K=1,NMAT
              UTMP2(K,K0)=UTMP1(K,K0+1)
    6125    CONTINUE
    6124    CONTINUE
            CALL SMAT(UTMP2,NMAT,CX1)
              CDH(M,I4,1)=1.0E0
            DO 6128 K=2,NMAT+1
              CDH(M,I4,K)=CX1(K-1)
    6128    CONTINUE
C
C
    6102    CONTINUE
    6101    CONTINUE
      74    CONTINUE
C
C
C           we now examine the Fourier coefficients
C           there is a GREAT deal of arbitrariness here;
C           fine-tuning may be necessary
C
            NII=MAG*(NA97)
            NFF=MAG*(LONG-NA97)
            NI=NII+2
            NF=NFF-2
C
C
C           giant do-loop follows:
C
            DO 7101 I=1,NTRIM
            NMAT=ITRIM(I)-1
            LKT=0
            LKTT=0
            DO 7102 JC=NI,NF-1
            DO 7601 N=-2,3
            DO 7602 K=2,NMAT+1
              CU(N,K)=CDH(JC+N,I,K)
    7602    CONTINUE
    7601    CONTINUE
            F40=CU(-2,4)+(1.0E0)-(CU(-2,2)**2)
            F41=CU(-1,4)+(1.0E0)-(CU(-1,2)**2)
            F42=CU(0,4)+(1.0E0)-(CU(0,2)**2)
            F43=CU(1,4)+(1.0E0)-(CU(1,2)**2)
            F44=CU(2,4)+(1.0E0)-(CU(2,2)**2)
            F45=CU(3,4)+(1.0E0)-(CU(3,2)**2)
            IF((F42*F43).LT.(0.0E0)) THEN
C           begin the detailed coefficient analysis
C           the end-statement is near 7102 (far below)
            C20=CU(0,2)
            C21=CU(1,2)
            C40=CU(0,4)
            C41=CU(1,4)
            A=C21-C20
            B=C41-C40
            E=B-(2.0E0)*C20*A
            E1=SQRT(E*E+(4.0E0)*F42*A*A)
```

```
          IF(F42.GT.(0.0E0)) THEN
            S=(E+E1)/(A*A*(2.0E0))
          ELSE
            S=(E-E1)/(A*A*(2.0E0))
          ENDIF
          SA=1.0E0-S
          RTEMP=SA*R4(JC)+S*R4(JC+1)
          DO 7603 K=2,NMAT+1
            BU(K)=SA*CU(0,K)+S*CU(1,K)
 7603     CONTINUE
          AY1=(CU(-1,2)*CU(-1,2)-(1.0E0))/CU(-1,4)
          A2=(CU(0,2)*CU(0,2)-(1.0E0))/CU(0,4)
          A3=(CU(1,2)*CU(1,2)-(1.0E0))/CU(1,4)
          A4=(CU(2,2)*CU(2,2)-(1.0E0))/CU(2,4)
          AH=(BU(2)*BU(2)-(1.0E0))/BU(4)
          B1=CU(-1,2)*CU(-1,3)/CU(-1,6)
          B2=CU(0,2)*CU(0,3)/CU(0,6)
          B3=CU(1,2)*CU(1,3)/CU(1,6)
          B4=CU(2,2)*CU(2,3)/CU(2,6)
          BH=BU(2)*BU(3)/BU(6)
 C
          Z=1.0E0
          AY1=ABS(AY1-Z)
          A2=ABS(A2-Z)
          A3=ABS(A3-Z)
          A4=ABS(A4-Z)
          AH=ABS(AH-Z)
          B1=ABS(B1-Z)
          B2=ABS(B2-Z)
          B3=ABS(B3-Z)
          B4=ABS(B4-Z)
          BH=ABS(BH-Z)
 C
          Q1=AMIN1(A2,AH,A3)
          Q2=AMIN1(B2,BH,B3)
          Q91=AMIN1(AY1,A2,AH,A3,A4)
          Q92=AMIN1(B1,B2,BH,B3,B4)
 C
          IF(Q91.GT.(.10E0)) THEN
          GOTO 7102
          ENDIF
 C
          IF(Q92.GT.(.10E0)) THEN
          GOTO 7102
          ENDIF
 C
          LKT=LKT+1
          RZ(LKT,I)=RTEMP
 C
 C        use LOCAL transversality to trim the list
 C
          FG=F43-F42
          C1=(F41-F40)/FG
          C2=(F42-F41)/FG
          C4=(F44-F43)/FG
          C5=(F45-F44)/FG
          Z89=AMIN1(C1,C2,C4,C5)
          IF(Z89.LT.(0.0E0)) THEN
 C        look at B2,BH,B3 versus B1,B4
            QDE=Q2-Q92
            IF(QDE.GT.(0.0E0)) THEN
            GOTO 7102
            ENDIF
            IF(Q2.GT.(.005E0)) THEN
            GOTO 7102
            ENDIF
          ENDIF
```

```
C
        IF(Q1.GT.(.05E0)) THEN
        GOTO 7102
        ENDIF
C
        IF(Q2.GT.(.05E0)) THEN
        GOTO 7102
        ENDIF
C
        LKTT-LKTT+1
C
C
        F61-CU(-1,2)*CU(-1,3)-CU(-1,6)
        F62-CU(0,2)*CU(0,3)-CU(0,6)
        F63-CU(1,2)*CU(1,3)-CU(1,6)
        F64-CU(2,2)*CU(2,3)-CU(2,6)
        V41-(F42-F41)/H4
        V42-(F43-F42)/H4
        V43-(F44-F43)/H4
        V61-(F62-F61)/H4
        V62-(F63-F62)/H4
        V63-(F64-F63)/H4
        AT1-ABS(V42-V41)
        AT2-ABS(V42-V43)
        AT3-ABS(V42)
        AT4-ABS(V62-V61)
        AT5-ABS(V62-V63)
        AT6-ABS(V62)
        R4T(LKTT,I)-AMAX1(AT1,AT2)/AT3
        R6T(LKTT,I)-AMAX1(AT4,AT5)/AT6
C
C
C       minimize the Hecke index next;
C       use full SMAT calls for CDHH
C       NB: very noisy cases can be made WORSE here!!!
C
C
        U27-FLOAT(MAG2)+(1.0E-6)
        MRG-INT(U27/(3.0E0))+1
        H4H=(.5E0)*H4
C
C
        DO 7171 JX-0,NDEG
        R-RTEMP+FLOAT(2*JX-NDEG)*H4H
        CALL KBMAT(R,U)
        DO 7172 K0-1,NMAX
        DO 7173 K-1,NMAX
          GIANT(K,K0,JX)-U(K,K0)
7173    CONTINUE
7172    CONTINUE
7171    CONTINUE
C
C
        DO 7621 JBB-0,2*MRG
          R97(JBB)-RTEMP+FLOAT(JBB-MRG)*H5
          WHN-FLOAT(NA97)+(.5E0)
          SX-WHN+FLOAT(JBB-MRG)/FLOAT(MAG2)
          CALL FLAG(SX,UTMP1)
C       prepare an adjusted copy of UTMP1
C       the throw-away row should be the same as for CDH!!!
        DO 7631 K-1,NMAT
          UTMP2(K,NMAT+1)--UTMP1(K,1)
7631    CONTINUE
```

```
          DO 7633 KO-1,NMAT
          DO 7635 K-1,NMAT
          UTMP2(K,KO)-UTMP1(K,KO+1)
 7635     CONTINUE
 7633     CONTINUE
          CALL SMAT(UTMP2,NMAT,CX1)
          CDHH(JBB,1)-1.0EO
          DO 7637 K-2,NMAT+1
          CDHH(JBB,K)-CX1(K-1)
 7637     CONTINUE
          C2-CDHH(JBB,2)
          C3-CDHH(JBB,3)
          C4-CDHH(JBB,4)
          C6-CDHH(JBB,6)
          C8-CDHH(JBB,8)
          C9-CDHH(JBB,9)
          WTA-(C2*C2)-1.0EO
          WTB-C2*C3
          WTC-C2*((C2*C2)-(2.0EO))
          WTD-(C3*C3)-1.0EO
          WH1-ABS(-1.0EO+(WTA/C4))
          WH2-ABS(-1.0EO+(WTB/C6))
          WH3-ABS(-1.0EO+(WTC/C8))
          WH4-ABS(-1.0EO+(WTD/C9))
C         the functional WH is adjustable!!!
          WH(1+JBB)-WH1+WH2
 7621     CONTINUE
          M98-(2*MRG)+1
          JBBM-ISMIN(M98,WH,1)
          JBBM-JBBM-1
          RZT(LKTT,I)-R97(JBBM)
          DO 7640 K-1,NMAT+1
          CDZ(LKTT,I,K)-CDHH(JBBM,K)
 7640     CONTINUE
C
C         compute velocity fluctuations
C
          DO 7641 N--1,1
          DO 7642 K-2,NMAT+1
          VE(N,I,K)-(CU(N+1,K)-CU(N,K))/H4
 7642     CONTINUE
 7641     CONTINUE
          DO 7651 K-2,NMAT+1
          AT1-ABS(VE(0,I,K)-VE(-1,I,K))
          AT2-ABS(VE(0,I,K)-VE(1,I,K))
          AT3-ABS(VE(0,I,K))
          VDZ(LKTT,I,K)-AMAX1(AT1,AT2)
          VDX(LKTT,I,K)-AT3
          VDY(LKTT,I,K)-AMAX1(AT1,AT2)/AT3
 7651     CONTINUE
C
C
          ENDIF
C
C
 7102     CONTINUE
          LCOUNT(I)-LKT
          LKOUNT(I)-LKTT
 7101     CONTINUE
C
C         now do the final print-out
C
          PRINT*,'    '
          PRINT*,'BASIC PARAMETERS:'
          PRINT 90, R1,R2
```

```
          PRINT 91, H3
          PRINT 92, H4
          PRINT 921, H5,MRG
          PRINT 93, NDEG
          PRINT 94, R4(NI),R4(NF)
  90      FORMAT(1X,F15.10,2X,F15.10)
  91      FORMAT(1X,'H3-',F15.12)
  92      FORMAT(1X,'H4-',F15.12)
  921     FORMAT(1X,'H5-',F15.12,2X,'MRG-',I6)
  93      FORMAT(1X,'DEGREE-',I3)
  94      FORMAT(1X,'ACTIVE RANGE:',F15.10,2X,F15.10)
C
C
          PRINT*,'   '
          PRINT 9411, NR, NC
  9411    FORMAT(1X,'ROWS:',I3,3X,'COLS:',I3)
          PRINT 9412, YA(1), YA(NMAX)
  9412    FORMAT(1X,'Y RANGE:',F6.3,3X,F6.3)
C
C
          DO 8101 I-1,NTRIM
          NMAT-ITRIM(I)-1
          PRINT*,'   '
          PRINT 95, ITRIM(I)
  95      FORMAT('ITRIM-',I4)
          PRINT*,'MAXIMAL LIST (level H4):'
          PRINT 96, (RZ(L,I),L-1,LCOUNT(I))
  96      FORMAT(1X,'R-',F15.10)
C
C
          PRINT*,'TRANSVERSE LIST (level H5):'
          PRINT 961, (RZT(L,I),L-1,LKOUNT(I))
  961     FORMAT('TR-',F15.10)
C
C
          IF(IPRINT.NE.0) THEN
          DO 8201 L-1,LKOUNT(I)
          PRINT 962, RZT(L,I),R4T(L,I),R6T(L,I)
  962     FORMAT('FOR R-',F15.10,3X,E11.4,2X,E11.4)
          C2-CDZ(L,I,2)
          C3-CDZ(L,I,3)
          C4-CDZ(L,I,4)
          C5-CDZ(L,I,5)
          C6-CDZ(L,I,6)
          C7-CDZ(L,I,7)
          C8-CDZ(L,I,8)
          C9-CDZ(L,I,9)
          C10-CDZ(L,I,10)
          C12-CDZ(L,I,12)
          C14-CDZ(L,I,14)
          C15-CDZ(L,I,15)
          WTA-(C2*C2)-1.0E0
          WTB-C2*C3
          WTC-C2*((C2*C2)-(2.0E0))
          WTD-(C3*C3)-(1.0E0)
          WTAA-C2*C5
          WTBB-C3*C4
          WTCC-C2*C7
          WTDD-C3*C5
          HH4-WTA/C4
          HH6-WTB/C6
          HH8-WTC/C8
          HH9-WTD/C9
          HH10-WTAA/C10
```

```
         HH12-WTBB/C12
         HH14-WTCC/C14
         HH15-WTDD/C15
         PRINT 98, HH4,HH6,HH8,HH9
98       FORMAT(1X,'RATIOS:',F14.8,2X,F14.8,2X,F14.8,2X,F14.8)
         PRINT 98, HH10,HH12,HH14,HH15
         PRINT 99,(K,CDZ(L,I,K),VDZ(L,I,K),VDX(L,I,K),VDY(L,I,K),
     &   K-2,NMAT+1)
99       FORMAT(3X,'C(',I3,')-',2X,E22.15,4X,E11.4,2X,E11.4,2X,E11.4)
8201     CONTINUE
         ENDIF
8101     CONTINUE
         END
C
C
C
         SUBROUTINE KBMAT(R,U)
C        CRAY VERSION -- SINGLE PRECISION
         IMPLICIT REAL(A-H,P-Z)
         PARAMETER(NMAX-60,NPAX-60)
         DIMENSION XA(NMAX),YA(NMAX),XP(NMAX),YP(NMAX)
         DIMENSION U1(NMAX,NPAX)
         DIMENSION U2(NMAX,NPAX)
         DIMENSION U(NMAX,NPAX)
         COMMON /DH3/ XA,YA,XP,YP
         COMMON /DH4/ NN1R,NN1L,NN2,NN3,PI2
         S2-(5.0E0)/R
         T1-(1.0E0)/(3.0E0)
         T2-(2.0E0)/(3.0E0)
         S-AMIN1(T1,S2**T2)
         DO 6101 K-1,NMAX
         X1-XA(K)
         Y1-YA(K)
         X2-XP(K)
         Y2-YP(K)
CDIR$    NEXTSCALAR
         DO 100 N-NMAX,1,-1
            YN-(PI2)*FLOAT(N)*Y1
            C-(ABS(R-YN))/R
            IF(C.LE.S) THEN
               CALL KBES2(YN,R,RKBES)
            ELSE IF(YN.LT.R) THEN
               CALL KBES1(YN,R,RKBES)
            ELSE
               CALL KBES3(YN,R,RKBES)
            ENDIF
            U1(K,N)-SQRT(Y1)*RKBES
100         CONTINUE
         DO 110 N-NMAX,1,-1
            YN-(PI2)*FLOAT(N)*Y2
            C-(ABS(R-YN))/R
            IF(C.LE.S) THEN
               CALL KBES2(YN,R,RKBES)
            ELSE IF(YN.LT.R) THEN
               CALL KBES1(YN,R,RKBES)
            ELSE
               CALL KBES3(YN,R,RKBES)
            ENDIF
            U2(K,N)-SQRT(Y2)*RKBES
110      CONTINUE
6101     CONTINUE
```

```
          DO 6102 K-1,NMAX
             X1-XA(K)
             X2-XP(K)
             DO 200 N-1,NMAX
                ZN1-(PI2)*FLOAT(N)*X1
                ZN2-(PI2)*FLOAT(N)*X2
                U11-U1(K,N)
                U22-U2(K,N)
                U(K,N)-U11*COS(ZN1)-U22*COS(ZN2)
200          CONTINUE
6102      CONTINUE
          RETURN
          END
C
C
          SUBROUTINE KBES1(YN,R,RKBES)   ⌐
C
C
          SUBROUTINE KBES2(YN,R,RKBES)   ├─ As in Appendix A.
C
C
          SUBROUTINE KBES3(YN,R,RKBES)   ⌐
C
C
          SUBROUTINE FLAG(SX,SV)
          IMPLICIT REAL(A-H,P-Z)
          PARAMETER(NMAX-60,NPAX-60)
          PARAMETER(NDEG-11)
          PARAMETER(MAXDEG-51)
          DIMENSION GIANT(NMAX,NPAX,0:NDEG)
          DIMENSION SV(NMAX,NPAX)
          DIMENSION P(0:MAXDEG),D(0:MAXDEG)
          COMMON /DAH/ GIANT
          DO 2300 J-0,NDEG
             P(J)-1.0E0
             D(J)-1.0E0
             DO 2310  K8-0,NDEG
                IF(K8.NE.J) THEN
                P(J)-P(J)*(SX-FLOAT(K8))
                D(J)-D(J)*FLOAT(J-K8)
                ENDIF
2310         CONTINUE
2300      CONTINUE
          DO 3300 K0-1,NMAX
          DO 3310 K-1,NMAX
             SQQ-0.0E0
          DO 3320 JB-0,NDEG
             SQQ-SQQ+GIANT(K,K0,JB)*(P(JB)/D(JB))
3320      CONTINUE
             SV(K,K0)-SQQ
3310      CONTINUE
3300      CONTINUE
          RETURN
          END
C
C
          SUBROUTINE SMAT(U,N,C)
          PARAMETER(NMAX-60,NPAX-60)
          DIMENSION U(NMAX,NPAX+1), C(NPAX)
          DO 1000 M-1,N
             TEMP-0.0E0
             MAXI-0
```

```
         DO 1100 J=M,N
           IF(TEMP.LT.ABS(U(J,M))) THEN
           TEMP=ABS(U(J,M))
           MAXI=J
           ENDIF
1100     CONTINUE
*        swap rows if necessary
         IF(MAXI.NE.M) THEN
           DO 1200 J=1,N+1
           T=U(MAXI,J)
           U(MAXI,J)=U(M,J)
           U(M,J)=T
1200       CONTINUE
         ENDIF
         TEMP=U(M,M)
         U(M,M)=1.0E0
         DO 1300 J=M+1,N+1
           U(M,J)=U(M,J)/TEMP
1300     CONTINUE
         DO 1400 J=1,N
           IF(J.NE.M) THEN
             T=U(J,M)
             U(J,M)=0.0E0
             DO 1500 K=M+1,N+1
             U(J,K)=U(J,K)-(T*U(M,K))
1500         CONTINUE
           ENDIF
1400     CONTINUE
1000     CONTINUE
         DO 2001 M=1,N
           C(M)=U(M,N+1)
2001     CONTINUE
         RETURN
         END
```

School of Mathematics
University of Minnesota
Minneapolis, Minn. 55455

Editorial Information

To be published in the *Memoirs*, a paper must be correct, new, nontrivial, and significant. Further, it must be well written and of interest to a substantial number of mathematicians. Piecemeal results, such as an inconclusive step toward an unproved major theorem or a minor variation on a known result, are in general not acceptable for publication. *Transactions* Editors shall solicit and encourage publication of worthy papers. Papers appearing in *Memoirs* are generally longer than those appearing in *Transactions* with which it shares an editorial committee.

As of March 1, 1992, the backlog for this journal was approximately 9 volumes. This estimate is the result of dividing the number of manuscripts for this journal in the Providence office that have not yet gone to the printer on the above date by the average number of monographs per volume over the previous twelve months. (There are 6 volumes per year, each containing about 3 or 4 numbers.)

A Copyright Transfer Agreement is required before a paper will be published in this journal. By submitting a paper to this journal, authors certify that the manuscript has not been submitted to nor is it under consideration for publication by another journal, conference proceedings, or similar publication.

Information for Authors

Memoirs are printed by photo-offset from camera copy fully prepared by the author. This means that the finished book will look exactly like the copy submitted.

The paper must contain a *descriptive title* and an *abstract* that summarizes the article in language suitable for workers in the general field (algebra, analysis, etc.). The *descriptive title* should be short, but informative; useless or vague phrases such as "some remarks about" or "concerning" should be avoided. The *abstract* should be at least one complete sentence, and at most 300 words. Included with the footnotes to the paper, there should be the 1991 *Mathematics Subject Classification* representing the primary and secondary subjects of the article. This may be followed by a list of *key words and phrases* describing the subject matter of the article and taken from it. A list of the numbers may be found in the annual index of *Mathematical Reviews*, published with the December issue starting in 1990, as well as from the electronic service e-MATH **[telnet e-MATH.ams.com (or telnet 130.44.1.100)**. Login and password are e-**math**]. For journal abbreviations used in bibliographies, see the list of serials in the latest *Mathematical Reviews* annual index. Authors are encouraged to supply electronic addresses when available. These will be printed after the postal address at the end of each article.

Electronically-prepared manuscripts. The AMS encourages submission of electronically-prepared manuscripts in $\mathcal{A}_{\mathcal{M}}\mathcal{S}$-TEX or $\mathcal{A}_{\mathcal{M}}\mathcal{S}$-LATEX. To this end, the Society has prepared "preprint" style files, specifically the amsppt style of $\mathcal{A}_{\mathcal{M}}\mathcal{S}$-TEX and the amsart style of $\mathcal{A}_{\mathcal{M}}\mathcal{S}$-LATEX, which will simplify the work of authors and of the production staff. Those authors who make use of these style files from the beginning of the writing process will further reduce their own effort.

Guidelines for Preparing Electronic Manuscripts provide additional assistance and are available for use with either $\mathcal{AMS}$-TEX or $\mathcal{AMS}$-LATEX. Authors with FTP access may obtain these *Guidelines* from the Society's Internet node e-MATH.ams.com (130.44.1.100). For those without FTP access they can be obtained free of charge from the e-mail address guide-elec@math.ams.com (Internet) or from the Publications Department, P. O. Box 6248, Providence, RI 02940-6248. When requesting *Guidelines* please specify which version you want.

Electronic manuscripts should be sent to the Providence office only after the paper has been accepted for publication. Please send electronically prepared manuscript files via e-mail to pub-submit@math.ams.com (Internet) or on diskettes to the Publications Department address listed above. When submitting electronic manuscripts please be sure to include a message indicating in which publication the paper has been accepted.

For papers not prepared electronically, model paper may be obtained free of charge from the Editorial Department at the address below.

Two copies of the paper should be sent directly to the appropriate Editor and the author should keep one copy. At that time authors should indicate if the paper has been prepared using $\mathcal{AMS}$-TEX or $\mathcal{AMS}$-LATEX. The *Guide for Authors of Memoirs* gives detailed information on preparing papers for *Memoirs* and may be obtained free of charge from AMS, Editorial Department, P. O. Box 6248, Providence, RI 02940-6248. The *Manual for Authors of Mathematical Papers* should be consulted for symbols and style conventions. The *Manual* may be obtained free of charge from the e-mail address cust-serv@math.ams.com or from the Customer Services Department, at the address above.

Any inquiries concerning a paper that has been accepted for publication should be sent directly to the Editorial Department, American Mathematical Society, P. O. Box 6248, Providence, RI 02940-6248.

Recent Titles in This Series

(See the AMS catalogue for earlier titles)